The Design and Analysis of Longitudinal Studies

Their Role in the Measurement of Change

The Design and Analysis of Longitudinal Studies

Their Role in the Measurement of Change

HARVEY GOLDSTEIN
Institute of Education
University of London

1979

ACADEMIC PRESS
London · New York · San Francisco
A Subsidiary of Harcourt Brace Jovanovich, Publishers

ACADEMIC PRESS INC. (LONDON) LTD.
24 28 Oval Road,
London NW1 7DX

United States Edition published by
ACADEMIC PRESS INC.
111 Fifth Avenue
New York, New York 10003

Second Printing 1981

British Library Cataloguing in Publication Data

Goldstein, Harvey
The design and analysis of longitudinal studies.
1. Developmental psychology—Longitudinal studies
I. Title
155'.07'22 BF713 79–50104

ISBN 0–12–289580–0

Printed in Great Britain by
Page Bros (Norwich) Ltd, Mile
Cross Lane, Norwich, Norfolk.

Foreword

During the last two or three decades there has been a continuous debate, sometimes quite mistakenly acrimonious, on the relative merits of longitudinal and cross-sectional methods in the study of change. It should be obvious that the two approaches are complementary rather than competitive. Even to some scientists, however, the drama of confrontation has more appeal than the labour of working together.

Professor Goldstein is not one of these: his work has involved the most fruitful collaboration with biologists, physicians, educators and psychologists. He is one of the few statisticians with long and varied experience of longitudinal studies. When R. H. Whitehouse and I started the Harpenden Longitudinal Growth Study in 1948, the methods of analysis available for the type of data we were accumulating were almost entirely graphical and approximate. Harvey Goldstein joined us in the Institute of Child Health in 1964; algebraic models appeared, and soon a set of longitudinal growth programmes suited to the University of London Computer.

In 1958 the National Birthday Trust, of which Sir Alan Moncrieff, then Director of the Institute of Child Health, was an influential member, sponsored a national survey of perinatal mortality by studying a cohort of children consisting of all those born in a particular week of that year. The cohort was followed up at various subsequent ages, in what became known as the National Child Development Study, and Professor Goldstein looked after its statistical aspects, first in my department and later as head of the statistical section of the National Children's Bureau. Thus Professor Goldstein not only has the practical experience of an intensive longitudinal study of physical growth in a small number of subjects, but also of an extensive follow-up study of a large number of subjects in whom educational and social data were the main concern.

In addition Professor Goldstein played the predominant role in the planning and analysing of a large-scale cross-sectional study of growth undertaken by the Cuban Ministry of Health in collaboration with the Institute of Child Health's Department of Growth and Development; so he is no stranger to cross-sectional problems.

This book is the outcome of this wide experience. It combines the requisite theoretical knowledge with a severely practical approach to the problems encountered in human field studies and in the analyses of always imperfect human data. It is for me as great a professional as personal pleasure to recommend it to all statisticians who have to do with longitudinal data, and to all auxologists and students of child development who wish to understand the basic framework of their subject.

March 1979

J. M. Tanner
Department of Growth and Development
Institute of Child Health

Preface

In research on growth, development and change, longitudinal studies play a special part. They repeatedly measure the same individuals over time, and thus permit the estimation of relationships involving measurements made at different times or occasions. Whether these relationships are used to describe detailed patterns of individual change; to predict, from measurements at an earlier time, the values of other measurements at a later time; or to obtain a greater insight into underlying causal mechanisms, it is possible to estimate such relationships only by carrying out a longitudinal study. The study of change, however, is wider than just the study of such relationships, and a major aim of this book is to set out the theoretical and practical problems of longitudinal studies within the more general framework of studies concerned with understanding the processes of change. This is done in such a way that the methods may be understood readily by all those who work in this area. Because my own background lies in studies of the growth and development of children, much of the methodology will be illustrated with applications from that field. Nevertheless, the methodology itself is more widely applicable, for example to "panel" studies such as are often used in market research, or to experimental studies of animals or plants.

The justification for writing this book is that no other comprehensive systematic volume exists. Of the available books dealing with this topic, four of the more important are as follows. First, there is the volume edited by Harris (1963), which is concerned with aspects of the mathematical and statistical theory. Secondly, the book by Wall and Williams (1970) contains a brief history and deals with the general need for longitudinal studies. Thirdly the book by Goldfarb (1960) gives a useful description of the longitudinal method but contains little discussion of theoretical and conceptual problems and the tools for solving them. Finally, the volume edited by Nesselroade and Baltes (1979) presents some recent contributions to the subject, but it is largely concerned with the behavioural sciences and does not provide a fully comprehensive or systematic treatment. The number of papers dealing with longitudinal studies which have been published, together with the considerable practical experience which has been gained by many workers, now make it possible to attempt to order and unify this knowledge.

In writing this volume I have made the basic assumption that the essence of the mathematical and statistical concepts involved in this subject can be presented as logical concepts which are intelligible to the non-mathematician, and hence undue mathematical complexities have been avoided. An acquaintance with elementary ideas of statistical analysis, however, has been assumed, but where more unfamiliar topics are introduced, as in Chapter 2, they are fully explained. In addition, there are some technical appendices and full references to the more detailed statistical and mathematical treatment of various topics for those readers who wish to pursue them. Fortunately, many statistical concepts lend themselves readily to visual display, and diagrams have been given where they can be useful. Apart from mathematical technicalities, however, there are concepts which by their very nature are complex, and no attempt has been made to avoid these for the sole reason that they happen to raise difficult issues.

My thanks are due to the National Children's Bureau for the use of its facilities while I was working there and to Mia Kellmer Pringle, director of the Bureau, for her support and encouragement.

I am immensely grateful to the following friends and colleagues who very kindly read and criticised drafts of various chapters: Michael Healy, Michael Hills, Gerald Hoinville, Pepe Jordan, Roger Jowell, Brian Lienard, Ian Plewis, Beverely Rowe, Jim Tanner, Bill Wall and Felicity Willetts. I owe a particular debt to Michael Healy, Michael Hills and Jim Tanner who not only read every chapter but also helped to sustain my enthusiasm through a long period of gestation. Without their support and criticism this book would have been much the poorer. Those faults which remain are entirely my own. My wife, Barbara, deserves my thanks both for providing great moral support and for spending many hours turning my drafts into reasonable English prose. Lastly, my thanks are due to Shirley Freeman, Lesley Hansen, Field Lyons and Esther Mason for their excellent typing.

References

Goldfarb, W. (1960). "An Introduction to Longitudinal Analysis: The method of repeated observations from a fixed sample". Free Press, Glencoe, Illinois.

Harris, C. W. (ed.) (1963). "Problems in Measuring Change". University of Wisconsin Press, Madison.

Nesselroade, J. R. and Baltes, P. B. (eds.) (1979). "Longitudinal Methodology in the Study of Behaviour and Development", Academic Press, New York and London.

Wall, W. D. and Williams H. L. (1970). "Longitudinal Studies and the Social Sciences," Heinemann, London.

March 1979 Harvey Goldstein

Acknowledgements

The author gratefully acknowledges the permission of the following publishers to reproduce some of the illustrations.

Wolters Noordhof Publishing, Groningen
Taylor & Francis, London
Plenum Publishing Corporation, New York
Castlemead Publications, Hertford

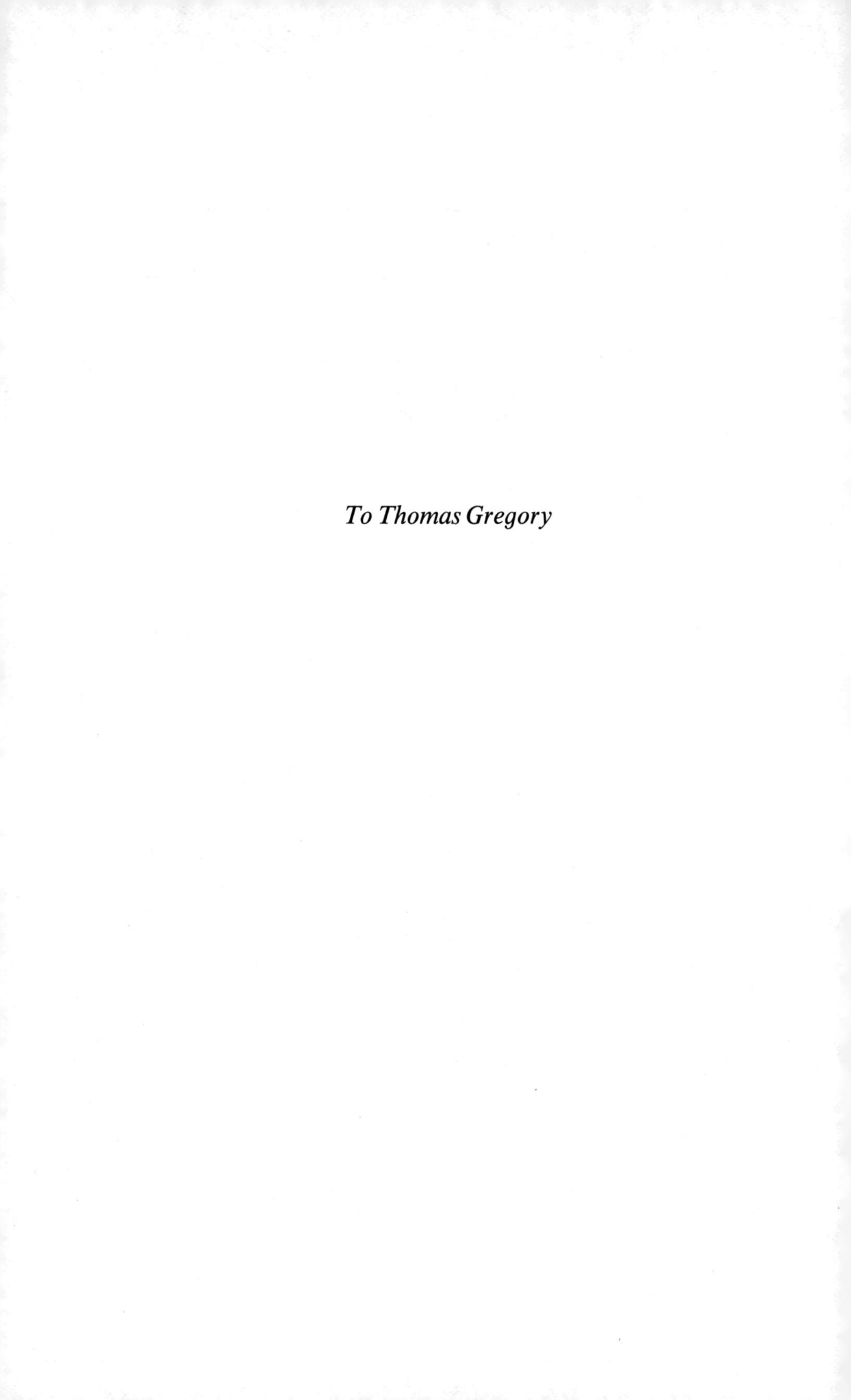

To Thomas Gregory

Contents

Chapter 3
The Choice of Measurement Scales over Time

Chapter 4
The Analysis of Models for Time Related Measurements

Chapter 5
The Analysis of Models for Relating Measurements at Different Occasions

Chapter 6
Population Standards or Norms

Chapter 7
Data Processing

Note on Statistical Notation

The notation and terminology used in this book follow closely the current practice among English speaking statisticians. In particular, population parameters are denoted by lower case Greek-letters and their sample estimates are denoted either by a circumflex over the corresponding Greek letter, or by the equivalent English letter. Upper case letters correspondingly are used for matrices. The commonly used letters are as follows:

Greek letter	*English letter*
α (alpha)	a
β (beta)	b
γ (gamma)	c
δ (delta)	d
ε (epsilon)	e
λ (lambda)	l
π (pi)	p
ρ (rho)	r
σ (sigma)	s
μ (mu)	m
χ (chi)	

In addition, the usual symbols $\bar{x}$, S are used for the sample mean and standard deviation.

Significance levels are denoted as follows:

***	$0{\cdot}001 > P$
**	$0{\cdot}01 > P > 0{\cdot}001$
*	$0{\cdot}05 > P > 0{\cdot}01$
Otherwise	$P > 0{\cdot}05$

1

The Theory and Practice of Longitudinal Studies

1.1 Definitions

In the widest possible sense longitudinal studies on human beings can be said to occur whenever the information collected, on all or some of the individuals in the study, relates to two or more points in time. Used in this sense such a definition would exclude very few studies of human populations although the longitudinal element might not be an important feature of very many.

Where information is collected at only one point, or "occasion", and information concerning previous times is obtained, say, by questioning or reference to files, such studies are called "retrospective". Thus the basic structure of a retrospective study is one where measurements are made only at the final occasion but where information is required for previous occasions. While we shall not discuss the special problems of retrospective studies except briefly in Chapter 5, it is important to recognise that they are not merely cheap substitutes for longitudinal studies as we define them below, but may often be the only practical way to collect the required data. For example, if information is required about the characteristics of mothers before a pregnancy began, this could not be obtained at the time without including the very much larger number of women who could possibly become pregnant in a specified period.

Where the design of a study relates to one occasion only, the term "cross-sectional" is used. Thus in a study of the body composition of five-year-olds, information might be collected about, say, previous morbidity in order to classify the children in subsequent analyses. This would not, however, alter the purpose of the study which is to make statements concerning children at one particular age.

The term "prospective" is used for studies where measurements are actually made at all occasions, and this will be used below to form the basis of a narrower definition of a longitudinal study.

The reader who is interested in a further discussion of the different kinds of study and an outline of the history of longitudinal studies, should consult Wall and Williams (1970) and Tanner (1979). It should be remembered that the important distinctions are not the semantic ones but those concerned with the questions asked by the study and with the quality of the information collected.

A longitudinal study in the narrow sense is one that contains individuals who are measured on more than one occasion. Embodied in this definition is the idea of a well-defined set of occasions. Even where data do not appear to fit the pattern of discrete occasions, we may be able to transform them in order to make them fit such a pattern. Consider the "event" data discussed briefly in Chapter 5, for example, where information is provided when an event such as a change of job occurs. By defining narrow time intervals and considering each one as an occasion, we can then describe the presence or absence of events in each one. Usually, however, an occasion is defined by a common time scale, for example age. In practice we are hardly ever lucky enough to be able to measure each individual at exactly the intended time though this need not prevent us from referring to the intended, or "target", occasions. The fact that measurements are actually *made* on every occasion creates a certain kind of time symmetry, the quality of the information obtained being similar at each occasion. In retrospectively collected information special problems arise because of the unreliability, often not quantifiable, in both the intended measurements themselves and in the exact times to which they refer. From this point onwards the term "longitudinal study" is interpreted in the narrow sense, as a study whose *principal aim* involves the actual measurement of individuals at several well defined occasions. In this sense it includes the so-called "pure" longitudinal study, where it is planned that every individual should be measured on every single occasion, and also the "mixed" study, where not all individuals are designed to be measured on all occasions. In the social science literature the term "panel study" is often used instead of "longitudinal study".

1.2 The Stages of a Longitudinal Study

1.2.1 Sampling

For some problems a "mixed" design may be preferred to a pure one, for example if we wish to obtain information over a ten year age interval and are unwilling or unable to follow the same sample over a ten year period. In this case we might choose three samples, each followed for four years, so that the age ranges overlap, thus allowing for some longitudinal analysis over the whole age range. The choice between different designs of this and other kinds is discussed in detail in Chapter 2.

In studying a sample drawn from a population, we are ultimately interested in making statements about the population. We are not interested in the particular group of five-year-olds we have measured, but all five-year-olds. However, the statement "all five-year-olds" is too vague unless positioned in a particular place at a particular time; "all five-year-olds in Paris in 1971" precisely defines a population. We can, of course, limit this population further by specifying the values of other variables, like sex, which help to define it, but the place, date and age-range are basic. In Chapter 2 we shall develop this notion further.

One of the major problems of longitudinal studies, and a feature which distinguishes them from other studies of populations, is the unscheduled sample attrition with the passage of time. This loss of individuals from a study requires special measures to ensure that biases do not arise to affect analyses and also often dictates the kind of sample which can be chosen, since not all individuals are equally likely to be lost. It may, for example, become worthwhile to choose those who are most likely to be retained, for example children with less mobile parents. The choice between different schemes is discussed in Chapter 2, and is shown to depend, like so many choices, primarily on the questions to be answered.

1.2.2 Measurement

Important as they are, the problems of defining and carrying out accurate and meaningful measurements will be dealt with in this book only in the most general way. The measurements themselves are peculiar to the particular field of study and, for the most part, are adequately covered in specialist works. It is the problems of controlling and interpreting such measurements when they are used in longitudinal studies which will be discussed.

We shall use the term "measurement" to mean any kind of well-defined method of collecting information. This may be the familiar type of measurement made with a ruler or balance, but also could be, for example, the categorisation of a subject's response to an attitude question in a questionnaire. The different types of measurement scales are discussed in Chapter 3.

1.2.3 Statistical Models and Analysis

The analysis stage may be dichotomized on the basis of the type of question to be answered. One type of question tends to be specific, being concerned with particular events or relationships. The other type involves exploring the data for interesting patterns. An example of the latter is the fitting of various types of growth curve which attempt to summarise growth patterns. There is a long history of this activity, including efforts to provide both a biological justification for observed growth patterns and to "graduate" growth by using relatively simple growth curves. The former type of question is concerned with such problems as the prediction of variables from one occasion to another, with the setting up of growth standards and with comparisons of growth characteristics from one population to another.

Since almost any mathematical manipulation of measurements assumes the existence of an underlying model, it is important to make that model quite explicit. Chapters 4 and 5 contain a more detailed discussion of the various models and criteria for evaluating them, but the following introduction embodies the ideas which are used throughout the book.

Suppose that we measure the thickness of a child's skinfold (at a given site); if we then talk about the "rate of growth" of the skinfold thickness from one age to another, we are tacitly referring to a mathematical model which states that the skinfold thickness is "linearly" (i.e. along a straight line) related to time. We may write down (with some elementary algebra) an equation which represents the relation between the skinfold thickness and age and which may be used to estimate this rate of growth as a single number, say b mm a year. Such a number is then a characteristic of the child on whom the original measurements were made and may be useful, for example, in categorizing the child as having a fast or slow rate of growth, etc.

Whether a particular model is "adequate", for example whether the categorization is accurate enough, is a question which can be answered only by studying the application of the model to data to see "how well" it works. If it happened that the skinfold thickness began to increase after the first measurement and then decreased to give the same value at the final measuring occasion, then the linear model would be an inaccurate summary since it would average an initial positive followed by a subsequent negative growth rate. A more appropriate model would be a "non-linear" one which allowed for the initial rise and subsequent fall.

1.2.4 Time Scales

Fundamental to the analysis of change is the problem of finding a suitable time scale. The most usual ones, historical time or chronological age, are suitable only if the events and patterns of change in each individual exhibit similar relationships to the time scale. For example, during adolescent growth, normal children exhibit similar patterns of growth in stature; that is to say, their growth curves have a similar shape. They are located, however, at different points on the age scale, with some children entering this phase of growth at an earlier age than others. Moreover, the time taken to complete this phase of growth varies. Thus, each child may be viewed as defining his own time scale. In this case we do not have enough information to say at the outset precisely what this time scale is, and it can usually be known only after the child has completely passed through this phase of growth.

Various attempts have been made to deal with this problem. One of these involves the construction of a "maturity" scale, such as that of "bone age", which is defined only in terms of relative biological maturity and which may, therefore, provide a more suitable universal scale for studying growth. Chapter 3 will discuss these problems.

1.2.5 Developmental Standards or Norms

In the study of change and development a common question is the following: is the measurement made on this individual at this age outside the limits within which most individuals' measurements fall? Clearly, the exact definition of the "reference population" of all individuals is of great importance. Are we referring, for example, to all seven year-old boys, or to all English seven-year-olds? The choice of a reference or sub-population, the selection of a sampling scheme and the most appropriate type of standard are discussed in Chapter 6.

1.2.6 Data Processing

In order to analyse a longitudinal study we need access to computing facilities. In Chapter 7 there is a discussion of the features necessary in a set of computer programs for this task. The problems which arise, however, can be summarized as follows.

The notion of a "target occasion" has already been introduced, and we may think of the information on a single individual as consisting of a set of examinations, each of which is connected with a particular target occasion. Computer terminology for such a set of examinations is a "record". If all individuals have exactly the same record structure, that is, are measured on the same occasions, then a relatively simple computer program can be written to deal with this structure. Unfortunately, however, individuals miss occasions and do not always come at the target times so that procedures have to be developed for recognizing and manipulating different structures.

It is convenient to divide the data processing into two parts. The first is concerned with manipulating the individual record structures into a standard form, and the second is concerned with statistical analysis of this standard data structure. That this distinction is useful both conceptually and practically can be illustrated by considering the problem of individuals who are not measured at a target time. They may, for example, be children who came a few weeks late for an appointment, or who did not come at all on an occasion. Since, at the analysis stage, it is convenient usually to base calculations on the assumption that all measurements are made at exactly the target times, for example, on children's birthdays, the first stage of the processing may be used to "adjust" measurements to the target times. Also at this stage any mistakes can be noted. The first stage is completed once any mistakes have been corrected and a standard record structure created for each individual. As the study proceeds these records are "updated" with further measurements. In addition the data processing system needs to be able to handle data which arise as "events", either by manipulating them into the form of a set of occasions or by providing a record of their occurrence in a readily available form. Finally, we can analyse this new well-defined record structure using the models described in Chapters 4 and 5.

1.3 Practical Considerations in Longitudinal Surveys

A longitudinal study where the researcher has control over the participation of his subjects, such as for example in a laboratory animal experiment, will have various practical problems to contend with. These will tend to be, however, the same kinds of problems which arise in studies which are similar but not longitudinal. A field of application will have procedures for coping with such problems, and we shall not pursue them here.

In research on human populations, however, there are relatively few systematic accounts of the practicalities of carrying out such surveys, and in the case of longitudinal surveys there are just a few references in rather widely scattered articles. The purpose of the present section is to draw together some of the experiences of those who have worked with longitudinal surveys and to set out the practical considerations which are particularly relevant to the collection of longitudinal information. For an excellent account of the detailed practice of survey research the reader is referred to the book by Hoinville and Jowell (1978) which can be read in conjunction with this section, and also to the report by Bailer and Lanphier (1978).

Biases in longitudinal studies tend to arise from problems such as refusals to cooperate, and will not be dealt with in this section but in Chapter 2 on sampling and design.

1.3.1 Overall Planning and Organization

In a single occasion cross-sectional survey it is normally possible to plan fully the various stages from the outset. The appointment of staff, the design and piloting of questionnaires or other measuring instruments, the training and deployment of field workers, the coding, checking, analysis and reporting of results, are familiar operations which many survey workers have had experience of organizing. In the case of longitudinal surveys, however, not only is there far less practical experience available, but also some quite new problems arise, so that a straightforward extension of the procedures appropriate to a cross-sectional survey is not appropriate.

The first and most obvious problem is that of coordinating the overlap of stages. Whereas, in a cross-sectional survey, the various procedures occur more or less sequentially, in a longitudinal survey it may be necessary, for example, to contact subjects while data from previous occasions are still being analysed. The design for a longitudinal survey will usually specify when occasions are to occur, so that it becomes of primary importance to collect data at these times, with the result that other activities may be curtailed. It is quite possible, therefore, to continue collecting data while having a backlog, not only of uncompleted analyses, but also of uncompleted coding, checking and file updating procedures. Moreover, in a longitudinal survey such uncompleted tasks tend to accumulate over time and it may become increasingly difficult to complete them satisfactorily even if the time and resources are available. Ideally, then, the complete strategy for a longitudinal survey needs to be clearly specified at the outset. However, because even the best laid plans can go wrong, this specification should allow for modifications to be made with the minimum of disruption. For example, it may be possible to postpone measurement occasions for some subjects, or to reduce the sample size etc. without too much loss of information.

Another important consideration with long-term studies, is that it is often impossible at the outset to foresee precisely what future information will be required, either for longitudinal or cross-sectional purposes. Theoretical developments may dictate the collection of unanticipated data, social and other policy aims may change, and even the early results from a study may indicate that new data should be collected or the time-sampling scheme should be modified. One solution to these problems is to anticipate as many possibilities as one can and to include some redundancy in the initial data gathered so that there is less chance to regret the failure to collect data where its relevance may become apparent only later. This, of course, needs to be tempered by the constraints of size. Crider *et al.* (1973) discuss this problem in some detail.

In Chapter 2 we deal with the technical problems of sampling and indicate how to achieve the most efficient designs. Prior to such considerations, however, come decisions about the purpose and content of a study, for these will limit the number of possible planning strategies. For example, it may be specified that occasions are to be a year apart in a study of physical growth, or, say four years apart in a study of children's educational development. Also, it is rare that a longitudinal study will be interested solely in longitudinal analyses, and some importance will attach to providing analyses of cross-sectional data. As pointed out in Chapter 2, it is often statistically more efficient to adopt a "mixed" rather than a "pure" longitudinal design and it will often provide more useful substantive information. Thus, an educational survey might require yearly measurements on a subsample of children supplemented by new and large cross-sectional samples at particular ages, such as those when children transfer between types of school. It is unfortunate that so little experimentation with these mixed designs seems to have taken place.

It is worth emphasizing the importance of planning in order to produce results on time. With a well organized data processing system, such as described in Chapter 7, it should be possible to carry out both cross-sectional and longitudinal analyses of data soon after any new data are collected. Apart from the obvious advantages of producing up-to-date information, the production of results is also important for the morale of the staff.

For all these reasons and for those given below it is apparent that considerable extra attention needs to be paid to organizational aspects of longitudinal studies from the outset, and this implies extra resources. Such extra resources may be a necessary condition for a successful study.

1.3.2 Tracing and Gaining Cooperation

It is clearly essential that a longitudinal study should be able to locate individual subjects when they are required for measurement. Methods for doing this efficiently will vary from study to study and from place to place. In some areas or countries, for example, centralized registers of individuals may be kept for health and welfare purposes, and these may be used to locate given subjects. In other areas population mobility may be very low, so that the effort allocated to tracing may be small. The following remarks will therefore need to be read bearing in mind the circumstances of any particular study.

Several authors (see, for example, Eckland, 1968) have suggested that, given sufficient resources, nearly all the subjects originally sampled can be traced, even after several years have elapsed. This is particularly the case for special groups who may be conscientious, for example parents of adopted children (Skeels and Skodak, 1965). In practice, however, resources will not be unlimited and the problem is to know how many resources are sufficient to achieve a satisfactory rate of tracing. The question as to what constitutes a satisfactory tracing rate is not an easy one to answer, depending in particular on how atypical the untraced individuals are likely to be. In many studies, failure to trace is associated with high geographical mobility so that the untraced subjects may well possess relatively unusual characteristics. Many would regard a tracing rate of below 90% as poor, and a reasonable target would be to trace at least 95% of the original sample at each subsequent occasion. Clarridge *et al.* (1978) show that in one study the time spent tracing an individual increases approximately linearly up to about 70% traced and after 80% the increase tends to become exponentially related to the percentage traced. If we assume that the costs of tracing also follow this pattern, then a judgement will be required as to when sufficient tracing has been done to reduce any bias to an acceptable level within reasonable time and budget constraints. There seems to have been little systematic attempt to quantify this problem.

The most efficient use of resources in tracing, as in other aspects of sample maintenance, is to begin with the method which uses least resources per individual. This may involve correspondence or a telephone call to the last known address, a search through a population register or an approach to institutions such as schools. Very often this will locate the majority of subjects, especially in studies where the occasions are close together. The next stage usually involves pursuing individuals using whatever methods are available. One useful method is to visit the last known address and ask the occupants, neighbours, etc., for information they may possess. If information about a subject's relatives has been obtained at the first occasion, then it may be possible to contact one or more of these to establish the location of the subject. The importance of obtaining such data as well as detailed identification and family information on a subject at the first occasion cannot be over-emphasized. Another approach which has sometimes been used successfully is to persuade the subjects themselves to notify the study whenever they move. Crider *et al.* (1971) and McAllister *et al.* (1973) give a detailed account of different methods of tracing subjects in a longitudinal study.

Once an individual who still qualifies as a member of the sample has been traced, we will seek to obtain measurements on him. At this stage we may be met with a refusal or a failure to make contact; problems which occur in all types of surveys. Where there is an initial failure to make contact we can try several times until contact is made. In interview surveys at least three times will be necessary usually to obtain a satisfactory response rate, and if this is expensive then we may choose to do it on a random sub-example only. Using a sub-sample at least enables us to estimate the extent of any biases which may be caused by the non-responders. Other methods have been suggested and the interested reader should refer to Kish (1965).

Refusals to cooperate have to be dealt with at a much earlier stage of a longitudinal study. Subjects or their families who have an interest in the study are less likely to refuse to cooperate and various methods of maintaining interest have been used. Financial incentives are often successful, and are justified by the time a subject devotes to the study. With some studies there may be a "spin off" for a subject in terms of, for example, regular medical examinations. The feedback of the results of a study to all concerned by means of individual letters and mass media publicity, is another valuable way of maintaining interest and contact, and a clear explanation of the purpose of the study is important. All this needs to be coupled with an honest explanation of the objectives and procedures of the study. Resources allocated to these aspects will usually be a worthwhile expenditure.

1.3.3 Piloting and Quality Control

In section 1.3.1 we discussed the need to avoid the several stages of a longitudinal survey interfering with each other and causing congestion. Piloting of measuring instruments such as questionnaires will normally be necessary before their use on the sample, although for simple repetitive studies, such as certain kinds of physical growth studies, piloting may only be necessary at infrequent intervals. Like the main sample, the pilot should be given to subjects who have been measured at previous occasions, so that the longitudinal information to be collected can be fully evaluated. One of the difficulties here is that, if a special longitudinal group is set aside for piloting, the measurements made on them at previous occasions will not necessarily be the same as those made on the main sample owing to modifications made after the previous pilots had taken place. Thus strict comparability may not be possible. One method of avoiding this is to plan to set aside, at each occasion, a small number of subjects from the main study, randomly chosen, to be used as a pilot for following occasions. At any given occasion we would, therefore, have a pilot group who had been measured at the previous occasion together with the pilot groups from all prior occasions. These groups would all be followed, but would have differing amounts of information comparable with the main sample. Once the size of the pilot groups has been decided, an appropriate increase in the overall main sample size can be made to cope with the losses expected as a result of extracting subjects for piloting. Cross-sectional pilot groups can also be used to evaluate the purely cross-sectional data, although these will not normally be able fully to replace the longitudinal pilots.

When measurements are made, for example, by skilled anthropometrists or by educational psychologists, or interviews conducted by skilled interviewers, training has to take place to ensure consistency and interpretability of measurements. Each subject area discipline may have its own set of procedures but these are often limited to training for a relatively short time period. In a longitudinal study, a subject may have measurements made by different people at each occasion, and it becomes very important to introduce a quality control procedure to ensure comparability. Even where the same measurer is used at each occasion, he or she may become more or less skilful or vary for some other reasons. Since absolute reference standards are rare, it is particularly important to provide very full documentation giving details of measurement procedures, supplemented possibly by visual recordings. Jordan *et al.* (1975) give an example of an elaborate quality control procedure for a physical growth study.

1.3.4 Checking, Coding and Documentation

We have already referred to the need for careful documentation in a long-term longitudinal study. In a cross-sectional study where a single research team is employed for the duration, it *may* be possible to avoid very detailed documentation of procedures, where these are well understood by all concerned. In a longitudinal study, not only do personnel change, but memories are fallible and detailed documentation of all procedures becomes extremely important.

The usual procedures for checking data are based on expected ranges or values of variables or on expected logical or statistical relationships between variables. Methods of doing this for longitudinal data are described in Chapter 7, but here we need to note that the rules for identifying "atypical" or "suspicious" values should be explicitly recorded. In addition, a record needs to be kept of all decisions made with regard to these values and on what general grounds. Incorrect decisions are always possible, and if details are recorded, subsequent data may provide a means of correction.

It is good survey practice to provide a codebook where all the data collected are listed, together with the response categories, variable ranges and any special features. During the process of coding data, for example when assigning individuals to categories based on an open-ended attitude question, all kinds of decisions are made by coders. As far as possible these should be systematized and written down and records kept for those cases where largely subjective judgements were necessary.

1.3.5 Staffing

As well as losing subjects, a long-term study is also prone to lose its research workers. It requires a considerable commitment to remain with one project for ten or 20 years, although there are several examples where this has occurred. It becomes important, therefore, to formalize the structure of a study to a greater extent than is usual with cross-sectional studies. An institutional commitment to a study is usually desirable so that accumulated experience can be utilized, consistent procedures can be maintained and the various items of documentation preserved. A career structure is important if it can be maintained, at least for a small "core" of staff. Staff commitment is likely to be enhanced if regular analysis of the data is encouraged, including past as well as current data.

Staff do change, however, and this should be recognized by the provision of adequate training and familiarization procedures for new staff, together with a recognition that they may bring new and useful ideas also. Changes in staff are also likely to lead to a greater mixture of disciplines than is common in cross-sectional studies, and considerable skill is involved in maintaining mutual understanding between disciplines.

1.3.6 Confidentiality and Informed Consent

There is now wide agreement on the need to preserve the confidentiality of the data obtained in surveys, and to obtain the freely given consent of subjects to participate. The problems of doing so, however, are not always easy to solve, although various codes of practice are in operation. A discussion can be found in Hoinville and Jowell (1978) and Bailer and Lanphier (1978). We mention briefly just two issues which need to be considered further when carrying out a longitudinal survey.

The first issue is that of safeguarding anonymity. This is usually done by removing identifying information from the original data source, for example a questionnaire, and substituting a code number which also appears on a separate file containing identifying information. This latter file is then kept secure. In a longitudinal study, data from separate occasions need to be collated and possibly cross-checked, thus entailing identification of the information on a subject. This may also occur at the time of tracing in order perhaps to "match-up" subjects. Thus, throughout the study there is a continual need to identify the subject's data and this will make the task of maintaining confidentiality a more difficult one than that normally found in a cross-sectional survey, where sometimes it is possible physically to destroy all identifying information after the data have been stored as a computer file.

The second issue is that of informed consent. This means, roughly speaking, that the information provided by the subject is freely given and that if the subject, during the course of being measured or questioned, decides to withdraw from the survey, then he or she is entitled to do so and to have any data removed from the study files. In a longitudinal study it is possible that a subject may additionally request that previously given data be removed from the file. Even if this was felt to be a reasonable demand, it is not clear that it could be met entirely satisfactorily since, in particular, the data for previous occasions may already have been analysed and the results published. Fortunately such cases seem to be rare so that removal of the data for these subjects would not affect appreciably any further analyses.

References

Bailer, B. A. and Lanphier, C. M. (1978). "Development of Survey Methods to Assess Survey Practices" American Statistical Association, Washington D.C.

Clarridge, B. R., Sheehy, L. L. and Hauser, T. S. (1978). Tracing members of a panel: A 17-year follow-up, *In* "Sociological Methodology 1978" Jossey-Bass, London.

Crider, D. M., Willits, F. K. and Bealer, R. C. (1971). Tracking respondents in longitudinal surveys, *Public Opinion Quarterly*, **35**, 613–620.

Crider, D. M., Willits, F. K. and Bealer, R. C. (1973). Panel studies: some practical problems, *Sociological Methods and Research*, **2**, 3–19.

Eckland, B. K. (1968). Retrieving mobile cases in longitudinal surveys, *Public Opinion Quarterly*, **32**, 51–64.

Hoinville, G. and Jowell, R. (1978). "Survey Research Practice" Heinemann Educational Books, London.

Jordan, J., Ruben, M., Hernandez, J., Bebelagua, A., Tanner, J. M. and Goldstein, H. (1975). The 1972 Cuban National Growth Study as an example of population health monitering: design and methods, *Annals of Human Biology*, **2**, 153–171.

Kish, L. (1965). "Survey Sampling" Wiley, New York.

McAllister, R. J., Goe, S. J. and Butler, E. W. (1973). Tracking respondents in longitudinal surveys: some preliminary considerations, *Public Opinion Quarterly*, **37**, 413–416.

Skeels, H. M. and Skodak, M. (1965). Techniques for a high-yield follow-up study in the field, U.S. Public Health Reports, **80**, 249–257.

Tanner, J. M. (1979). "A History of the Study of Human Growth" Cambridge University Press, Cambridge.

Wall, W. D. and Williams, H. C. (1970). "Longitudinal Studies and the Social Sciences" Heinemann, London.

2
Sampling and Design

Over and above the usual problems of human population sampling, studies of change need to consider how to choose sample units from different occasions. The first part of the chapter deals with populations and sampling strategies, and is followed by a section on problems of design and inference in the measurement of change.

2.1 Defining a Population

It is well known that the characteristics of a geographically defined population change over time. For example, the mean height of British seven-year-olds increased by about 1·5 cm per decade in the first half of the twentieth century and this "secular trend" also affected the ages of occurrence of particular events such as menarche (Tanner, 1955). This historical process is therefore superimposed upon the changes which occur with age, and although both age and historical time are measured in the same units they are nevertheless logically distinct concepts. Hence, when defining a population, its historical origin needs to be specified, as well as its geographical location and its other defining characteristics.

Since any study of a human population takes place at a particular historical time, inferences about the population are strictly appropriate only to the time when the study was carried out. Hence we must assume some stability in the population being studied if we wish to make statements about the population at other times, past or future. For some characteristics of some populations such assumptions may be reasonable, particularly if supported by a series of studies at different times. With longitudinal studies, however, the period of the study itself, sometimes 20 years or more, may be long in relation to changes in the population. During this time the characteristics of the population being studied may have changed considerably, both in terms of average values, such as height at different ages, and also in terms of the relationships between variables. For example, a study of children from birth to the age of nine years may establish a relationship between standards of obstetrical care and physical growth. Subsequent changes, say in the availability of obstetrical care or in the demographic structure of the population, may considerably alter the nature of the relationship. For example, elective caesarian section in some populations is known to be confined largely to privileged groups. If, for this reason, it is also associated with faster growth, then changes in practice or policy which resulted in elective caesarian section being available to all sections of the population would tend to remove this association.

To attempt to deal with such a situation, two strategies are available. First, we can plan a series of similar studies starting, say, at three-year intervals, to determine how the average values and relationships change over time. Inevitably, however, so far as the relationships over time are concerned, the most recent study will be out of date, by as much as nine years in the above example. Hence, we are left with having to extrapolate the trends from our previous studies. It should be noted that this problem is especially important in a longitudinal study, and is of far less importance in a cross-sectional study where the only time delay is between collecting information and analysing it. The second strategy is to collect additional information which may suggest changing patterns. For example, by studying the rate of caesarian section, its distribution in the population and its relation to an immediate outcome of pregnancy such as birthweight, we might obtain indications of a changing relationship with subsequent growth.

Ultimately, however, our ability to utilize the results of an observational study will depend on our degree of understanding of the underlying causal relationships, and one function of observational studies is to advance our understanding of these. We discuss now a general framework which allows us to set a study within its historical context.

2.1.1 The Age–Time Plane

Figure 2.1 has two axes representing historical time and age, so that an individual who has a particular age at a particular time can be uniquely specified by a point. More generally we can specify whole populations in terms of their positions. For example, the area between the parallel lines at 45° to the axes marked I represents a "cohort" born around 1960 and studied at the ages indicated by the shaded areas labelled a_1–a_4 until 1975 when they were 15 years old. Likewise, the area marked II represents another cohort who were nine years old around 1960 and were studied then and at 15 years of age.

If we add a third perpendicular axis we can plot along this the mean value of a measurement such as weight, to give an age–time surface whose distance above the plane is the population mean weight. We can, in principle, study complex "multivariate" surfaces for many variables simultaneously and it is the exploration of the properties of such surfaces which is the aim of statistical analysis.

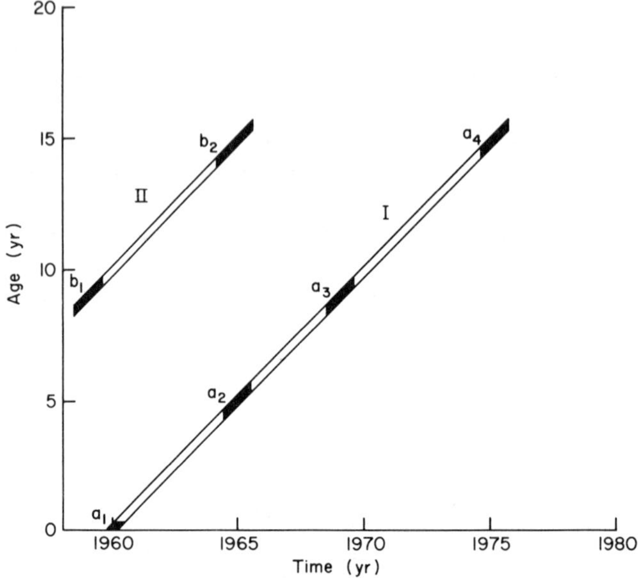

Fig. 2.1 Populations defined by age and time.

Various ways of specializing or simplifying the description of this age–time surface are available. We might, for example, assume that for a given age or ages no change in the average value of a variable occurred over time (these changes have already been referred to as "secular trends"). If this were the case, we could reasonably extrapolate findings in one time period to another time period. To study whether secular trends are occurring we would compare measurements made at the same ages at different points in time. For example, in Fig. 2.1 we might compare those measurements made at the occasions denoted by b_1 with those denoted by a_3 to see whether nine-year-olds in 1959 had the same characteristics as nine-year-olds in 1969. We might also wish to study whether the change in, say, mean weight between the ages of nine and 15 years was the same for cohort II as for cohort I.

It is worth emphasizing that the difference between the means of subgroups may be just as responsive to a secular trend as the mean values themselves. This is illustrated by the following example. Table 2.1 gives estimates of mean stature of 11-year-old British boys and girls from two national surveys, the National Survey of Health and Development (Douglas, 1964) and the National Child Development Study (Davie *et al.*, 1972).

Table 2.1 Mean height in cm at 11·0 years[a] (Standard errors in brackets)

	National Survey of Health and Development (1957)		National Child Development Study (1969)	
Boys	140·9	(0·154)	142·8	(0·097)
Girls	140·8	(0·174)	143·3	(0·108)
Boys-Girls	0·1	(0·235)	–0·5	(0·145)

[a] The values have been corrected to a mean age of 11·0 years using the height velocity standards of Tanner *et al.* (1966). The unpublished data from the National Survey of Health and Development are quoted by kind permission of Dr J. W. B. Douglas.

In the 12 years between the two surveys the mean value for boys has increased by 1·9 cm whereas that for girls has increased by 2·5 cm, the difference possibly being a result of earlier puberty which resulted in many of the girls, but not boys, already having entered their adolescent growth period by the age of 11 years. Thus we have an example of a secular trend operating to change both overall mean values and also the difference between subgroups of the population.

2.1.2 Cross-sectional and Longitudinal Designs

When sampling over a set of occasions we need to consider how much of the sample is "longitudinal", that is which individuals are measured on more than one occasion, and which occasions these are. As far as possible we shall approach this by assuming that the most efficient sampling scheme will have a longitudinal element, and then, in particular applications, determine the most efficient mixture of longitudinal and cross-sectional elements. Any study with a longitudinal element will be referred to as a "mixed longitudinal study" or more simply a "longitudinal study". Where every individual is measured at each occasion we shall use the term "pure longitudinal study". Likewise, where there is no longitudinal element, we shall use the term "cross-sectional study". Thus, in Fig. 2.1, measurement occasions denoted by a_2 and b_2 would constitute a cross-sectional study consisting of two age groups of five- and 15-year-olds in 1965. A pure longitudinal study would sample the same individuals belonging, say, to cohort I at each occasion.

We have already used the term "cohort" to describe a study in which all the individuals born within certain time limits are followed up. The two studies referred to above are both cohort studies consisting of all the births in Britain in a single week. We have the choice, when studying a cohort, of either sampling different individuals from the cohort at different ages or of following the same individuals. The importance of the distinction will become clear when we discuss how to obtain efficient estimates and how to deal with the problem of sample losses. For the moment we should note that if interest lies in developmental changes, then, whether or not the same individuals are used on different occasions, we are basically interested in age changes within well-defined cohorts. As we have already indicated, the presence of secular trends can distort the comparison of ages from different cohorts.

2.1.3 Superpopulations

So far we have used the word "population" rather loosely to describe a real population of individuals. Statisticians use the term to refer either to a finite or to a conceptually infinite set of individual units about which they wish to make inferences. Such inferences, for example, might be statements about subpopulation differences or mean values of measurements. In one sense a cohort study such as the British National Child Development Study is a finite population, defined by birth date. In practice, however, we would like to make statements not simply about a single week's births but about all children born at around that time. Because of possible secular trends, the vagueness of the phrase "around that time" is inevitable, but this lack of precise definition does not preclude us from hypothesizing a "superpopulation" from which our actual sample has been drawn. It is, however, a superpopulation with changing characteristics, and it is these changes to which the name "secular trends" really refers. In the study of development it is normal to conceive of the population of interest as infinite, and we shall assume in the remainder of this chapter that samples are selected from infinite populations.

2.1.4 Formal Designs

A brief word is necessary about various attempts which have been made to provide a formal model for different types of study design on the basis of the situation represented by Fig. 2.1. Schaie (1965) and some subsequent authors (e.g. Mason *et al.*, 1973) have suggested that three separate "effects" can be distinguished, namely age, time and cohort effects. As we have seen, a cohort can be defined by specifying appropriate times and ages. Likewise the scale of age may be defined by specifying appropriate times and cohorts, and time may similarly be defined in terms of ages and cohorts. It is clear that there are really only two underlying dimensions here (conveniently considered as age and time) and to attempt to classify research designs simultaneously in terms of all three, while conveying a certain "symmetry", only leads to circular and often misleading arguments. It also tends to complicate what is a relatively straightforward concept by attempting to link the different designs to hypotheses concerning genetic or cultural changes. The basic fallacy in the argument was pointed out by Baltes (1968), and it appears that much of the confusion on this issue centres around the failure to distinguish the "confounding" of three measurement scales which arise from the two-dimensional nature of the underlying classification, from the more usual "confounding" which can arise between logically independent variables within a particular experimental design. (See also Goldstein, 1979.)

2.1.5 Geographical Limits

To define a population in space at a particular point of time is in principle a straightforward matter, for example by using national or local boundaries. Thus we may define the population of all five-year-old children in a particular administrative region, and devise suitable methods to sample and measure them. If we wish to follow up the population, either by remeasuring the same individuals or new individuals at a subsequent time, we immediately face a new problem. Because of death and mobility the predefined population will change, acquiring new members and losing others. Hence the definition of the population must extend over time.

With respect to time we may therefore define the population in four different ways as follows:

(i) All those individuals who were living in the region at the start of the study, and who were still living in the region at each subsequent measuring occasion.
(ii) The above group together with all those at each subsequent measuring occasion who had come to live in the region after the first occasion.
(iii) All those who were living in the region at the first occasion, wherever they happened to live at subsequent occasions.
(iv) All those who were living in the region at the first occasion together with those who came to live in the region after the first occasion.

With each definition we are able to answer different questions and are faced with different practical problems. If we restrict ourselves to the first definition then the practical problems of follow-up will be easiest, but we have to face the possibility that the individuals who leave the region have different characteristics from those who stay, which results in biases. We note, however, that it is the population which suffers bias here and not, as is normally the case, just the sample. Indeed, we may be mainly interested in this biased population if the purpose of the study is to predict, say, those children who may require special services in the region at a subsequent age, so that it would not be appropriate to include those who move away. If, however, we wished to estimate changes in numbers of children requiring special services between the first and second occasion it would be appropriate to adopt the second definition.

Where we wish to answer questions about the change in individuals, then we require longitudinal information and it would seem appropriate here to use the third definition. If, for example, we wish to prepare longitudinal growth standards starting at a particular age then we should want to include all the children measured at this age. This would also allow us to compare the growth of those individuals who moved and those who stayed. The fourth definition encompasses the others and allows the most general comparisons to be made. It is also usually the most costly. The costs involved with definitions (iii) and (iv) clearly depend on local mobility factors which will vary greatly from country to country and from time to time. For example, Douglas and Blomfield (1958) found that in Britain in the late 1940's during the first four years of their lives, 53% of children's families had not moved at all, 31% had stayed within their local authority area and 14% had crossed the local authority boundary. Of those who crossed the local authority boundary a third had moved completely out of the region. They also showed that relatively more of the non-local moves were among families where the child's father had a non-manual occupation. Hence, in the light of known social class differences in child development, definition (i) would produce different average estimates of, say, growth in height from definition (iii).

2.2 Sampling Designs

Most of the theory of sample survey designs refers to a single sampling of a population. We begin, therefore, by outlining the main ideas behind the various designs and in subsequent sections see how these can be applied to longitudinal studies. More comprehensive accounts of sample survey theory can be found in Moser and Kalton (1971) or Kish (1965).

2.2.1 Simple Random Sampling

At the basis of sampling theory is the notion of a simple random sample. This is a selection of individuals from a defined population or sub-population where all samples of a given size have the same chance of selection, and in particular each individual member of the population has the same chance of selection. The choice of individuals is made using a set of random numbers, for example by numbering sequentially every member of the population and using a table of random numbers or a computer generated set of random numbers to select the sample members from this list. The basic sampling unit need not be an individual person but might be a household, school, etc. Simple random samples of whole populations are rare in surveys of human beings. For reasons to be outlined below, the population is often divided into subpopulations, for example geographically, within which random samples are then drawn.

An important reason for randomness lies in its protection against biased selection. One of the best known instances of a non-random selection which resulted in a biased selection is the Literary Digest telephone poll of US electors before the 1936 presidential election, enquiring about their voting intentions. With a sample of ten million individuals the wrong result was predicted. This happened, at least in part, because the sample was taken from the telephone directory which is clearly a non-random list of the total electorate. It is not just the mean value of a variable which may be biased because of non-random sampling, however, but also the dispersion or spread of values. For example, interviewers simply told to select random samples using subjective measures will often unconsciously choose "typical" individuals and avoid those with extreme characteristics.

If we wish to select a simple random sample from a large population we must first list all the individual units and then select a random sample of these. Such a list is called a sampling frame and typical examples are electoral registers and birth registers. Unfortunately, for many real populations of interest, such as very young children, there is little or no possibility of obtaining an accurate list of this kind. Where overall population lists do not exist we can adopt what are known as "multi-stage" sampling methods which involve compiling lists only for a few small subpopulations. Often, we have some pre-existing knowledge about the population in terms of what we are measuring, for example that the average height varies between geographical regions. In such a case we can use this information to improve the precision of the estimates of population means or proportions by "stratifying" the populations.

2.2.2 Stratified Sampling

Suppose we wish to estimate the average height of seven-year-olds in a population, measuring a sample of 1000. It is well known that a child's height varies inversely with the number of children in his household, and we may consider using household size as a stratifying factor. We might choose, say, five strata; households with one child, those with two, three, four, and five or more. We could do this by obtaining household size information from all seven-year-olds using, for example, school records. Table 2.2 shows a typical distribution of children in the resulting strata.

A common method of sampling within strata is to choose a random sample within each stratum whose size is proportional to the known number of all individuals within that stratum. The third column of Table 2.2 shows how many of the 1000 children would thus fall into each household group. If school records were available with household size information for each child, then we should be able to select random samples within each stratum from these records. This method of stratified random sampling is known as proportionate sampling and with it we ensure that each population unit has the same probability of being selected. It also has the desirable feature that

Table 2.2 Distribution of the Number of Children in the Household[a]

Number of Children in household	% of children in each family size stratum	Proportional allocation of sample numbers
1	8·9	89
2	34·8	348
3	26·0	260
4	15·1	151
5+	15·2	152
Total	100	1000

Variance within each stratum = 36 cm^2
Variance between all individuals = 42 cm^2
[a] Data obtained from Davie *et al.* (1972)

the best estimate of the mean height (or other variable) of the population turns out to be the same as if it were a simple random sample, namely the arithmetic mean of the individual measurements,

$$\bar{x} = \frac{1}{n} \sum_{i=1}^{k} n_i x_i = \frac{1}{n} \sum_{i=1}^{k} \sum_{j=1}^{n_i} x_{ij}$$

where x_{ij} is the height of the jth child in the ith stratum, n_i is the size of the ith stratum, x_i is the mean of the ith stratum and n is the total sample size.

The variance of $\bar{x}$, however, is not the same as for a simple random sample which is $\frac{\sigma^2}{n}$ where σ^2 is the between-individual variance. We have

$$\text{Variance } (\bar{x}) = \frac{1}{n^2} \sum_{i=1}^{k} n_i \sigma_i^2$$

where σ_i^2 is the variance between individuals within the ith stratum.

The ratio, variance $(\bar{x})/\left(\frac{\sigma^2}{n}\right)$ is known as the "design effect" (deft). If this is less than or equal to 1, as is always the case for proportionate sampling, then for a given sample size we obtain an estimate of the mean at least as precise as that for simple random sampling. This gain in precision comes from the use of the information available about strata differences. The design effect will be equal to one only if there is no difference between strata. Thus we are assured that when we come to make estimates for other variables, some of which may not be related to the stratification factor, we will not obtain a precision worse than that for a simple random sample.

In the example in Table 2.2 we estimate the gain in precision as follows. Using the estimates for the within strata variance and the variance between all individuals in the population we have,

$$\text{variance}\ (\bar{x}) = 0{\cdot}036\ \text{cm}^2;$$

$$\frac{\sigma^2}{n} = 0{\cdot}042\ \text{cm}^2;$$

$$\text{deft} = 0{\cdot}857.$$

Thus by using the stratified sample with only 857 individuals, we can achieve the same precision as a simple random sample of 1000 individuals. We shall return to this question of precision in a later section.

In the above example, if it were not feasible to obtain the household size information on each child prior to sampling, we could select a simple random sample and then apply what is known as "post-stratification". Having selected the sample we obtain the information about household size at the time of measuring in order to classify the children into the household size strata, and estimate the mean as

$$\bar{x} = \sum_{i=1}^{k} p_i x_i$$

where p_i is the population proportion of individuals in the ith stratum, which is assumed to be known from other sources *a priori*. This procedure uses the prior information on the population in a slightly different way and gives an estimate which has about the same precision as the ordinary stratification procedure.

Although proportionate sampling has many advantages, there are circumstances where different strata sampling fractions, that is the proportions of the total population in each stratum, are desirable. This is relevant when stratum measures are important in themselves. For example, the British National Survey of Health and Development followed up only 25% of the original cohort of children whose fathers had manual occupations. This was done in order to retain, for a fixed total sample size, sufficient numbers of children from non-manual backgrounds to provide reasonably precise estimates within that stratum. In other situations the cost of selecting one individual may differ between strata, so that in order to obtain the most precision for a given total cost, disproportionate sampling is required. The optimum sampling fractions are proportional to $\sigma_i/\sqrt{c_i}$ where c_i is the average sampling cost per individual in the ith stratum. Such a situation may arise when measurers have to travel from a centre so that with a geographical stratification the more distant regions will tend to be more expensive.

2.2.3 Multi-stage and Cluster Sampling

In stratified sampling we need to have an adequate sampling frame of the population and if this is unavailable, the cost of constructing one may be very large. To overcome this difficulty we can select in stages, for example by dividing a country into a large number of areas, selecting a sample of these and then forming a list of just the individuals in these areas. Such a procedure can also reduce the costs considerably by reducing the size of the sampling frame and concentrating field work in a few areas. We can add further stages if we wish, for example by sampling schools within each area and classes within schools. We would then only need to list the children within each selected class and make a selection from this list. In many circumstances it is convenient to select all the individuals within this final stage unit and this is known as "cluster sampling". The term "clustering" is used often, more generally when only a sample of the individuals in the final stage units is selected.

The penalty usually paid for cutting costs is a drop in precision, the design effect typically being greater than 1. This results from the fact that individuals within clusters are almost always more nearly alike than are all individuals in the population. This is a general attribute of geographically or institutionally defined clusters such as schools or classes.

The most important type of multi-stage sampling is where each population unit has the same probability of being selected so that, as with proportionate stratified sampling, we obtain a "self weighting" sample. One of the most widely used methods of achieving this is known as sampling with probability proportional to size (pps). A sample of the first stage or primary sampling units (psu) is randomly selected, each one having a probability of selection which is proportional to the number of individuals within it. For example if the psu's are geographical regions, census data can tell us the number of such individuals. This is continued until the same number of individuals are selected from each final stage unit. This procedure is often combined with a preliminary stratification, and it is often possible to arrange the relative amounts of stratification and multi-stage sampling to produce an overall design effect close to 1.

In some situations we do not know the exact sizes of the psu's, for example any census data may be out of date, but we may have *estimates* of their sizes. If we select with probability proportional to these estimates, we can make adjustments to the numbers of individuals sampled from the psu's when the size of these is known, following the listing of next stage units.

2.2.4 Complex Statistics from Complex Samples

So far we have discussed only the estimates of means or proportions using complex sampling procedures. We conclude this section with a few remarks on the use of more sophisticated statistical procedures, such as regression, where complex sampling has occurred.

Nearly all techniques commonly used in statistical analysis assume that a simple random sample of independent measurements is available. If we apply such techniques to data obtained from complex samples, treating them as if they were really simple random samples, significance tests will be distorted and we will obtain estimates of parameters which may be inconsistent. (An inconsistent estimate of a population value is one which does not approach the population value however large the sample on which it is based. A biased estimate need not necessarily be an inconsistent one.) The problem arises most severely with multi-stage samples where the clustering of individuals within the final stage units, as mentioned above, leads to a small positive correlation between the individuals.

Kish and Frankel (1974) have studied the effects of complex sampling in multiple regression and suggest that the usual estimates of the multiple regression coefficients (equally weighted if the sample is self-weighting) are consistent, and the design effect for the variance of each coefficient will usually be greater than 1 but less than the design effect for the mean value of the corresponding independent variable. This therefore allows us to correct the estimated standard errors for the coefficients using the design effects for their mean values, so giving a "conservative" confidence interval or significance test. The method these authors use to derive their results is called balanced repeated replications and it involves carrying out separate analyses on different portions of the total sample. It can be applied to any kind of complex sampling and in particular it may be applied to techniques for analysing longitudinal data which arise from complex samples. An alternative approach is to make explicit allowance for the sampling design in the statistical model. Where stratification has been used, constants can be fitted to express between-strata differences, and where multi-stage sampling has been used, then terms representing random differences between stage units can be incorporated into the model. Holt *et al.* (1978) develop models of this kind.

We shall deal with sampling for developmental standards in a later section, but it should be noted that the estimation of population percentiles using a complex sample can be carried out as if the sample were a simple random one. (Naturally, where the sample is not self-weighting then the individuals are given their appropriate weights.) Actually the estimates so obtained are slightly biased, although consistent. For example, if we use the sample standard deviation to estimate the percentiles for a normally distributed variable, the average or expected value of the usual sample estimate of variance is

$$\left\{\frac{n-d}{n-1}\right\}\sigma^2$$

where σ^2 is the true population variance and d is the design effect for the mean. Since d is rarely greater than 3, the bias is negligible when n is 100 or more, and as we shall see in a later section, the recommended minimum sample size for percentile estimation with a simple random sample is several hundred. For descriptions of developmental studies which use a complex sample see Malina *et al.* (1974) and Jordan *et al.* (1975).

When we come to estimate the standard errors of the percentile estimates the design effects may become more important. For details the reader should consult Kish (1965).

2.3 Sample Size Requirements

We have referred to the precision of estimates of mean values, and used the variance of these estimates as a criterion for comparing different sampling schemes. For an unbiased estimate (and for a consistent estimate in large samples) the smaller its variance the nearer it is likely to be to the "true" population value. More precisely we can define limits for the population value which should have a high probability of actually including the true value between them. The most common such limits are known as "confidence limits", and are described as follows.

If $\bar{x}$ is the mean of a random sample then 95% confidence limits, a,b are two values such that for samples selected in the same manner, the true population mean will lie between them in 95% of cases. If the distribution of the measurement is normal, then for a simple random sample a and b are given by

$$a = \bar{x} - 1{\cdot}96S/\sqrt{n}$$

$$b = \bar{x} + 1{\cdot}96S/\sqrt{n}.$$

Here S is the estimated standard deviation of the measurement, n is the sample size and hence $S/\sqrt{n}$ is the standard deviation or standard error of the sample mean. The value 1·96 is chosen because 95% of the normal distribution lies in the symmetrical interval $-1{\cdot}96$ to $+1{\cdot}96$ or approximately -2 to $+2$. (If n is less than about 30 we should use corresponding values from the "t" distribution rather than the normal.) We may express this formally as

$$\text{Probability } (a \leqslant \mu \leqslant b) = 0{\cdot}95$$

where μ is the true population mean.

Thus, with these limits we can be sure that there is a 95% chance that our confidence interval includes the mean. Of course, we can be more stringent by asking for, say, a 99% confidence interval but it will be wider than before, the corresponding normal distribution values being $-2{\cdot}33$ and $+2{\cdot}33$. Where the distribution is not normal or we are estimating a more complex statistic such as a ratio, then the method of calculating confidence limits will be different, although the principle is the same. The standard error of the sample estimate is not always used directly in the estimation of the confidence interval, but nevertheless remains a useful indicator of relative precision. Moreover, a useful feature of many estimates as the sample size becomes large is that we may regard them as having approximately normal distributions, so that the calculated standard error may be used to set confidence limits as above.

2.3.1 Significance Testing

Before going on to develop the above method for obtaining a specified precision, we need to mention an alternative approach which may be applicable in certain circumstances.

Suppose we wish to study the average difference between two subgroups of the population, for example the heights of children who have participated in one nutritional programme and the heights of a comparable group who have participated in another. We may wish to ensure that we can determine which of the two programmes resulted in a greater mean height. If we write $\bar{z}$ for the difference between the two means, then we may carry out a conventional significance test of the "null hypothesis" that the expected value of $\bar{z}$ is zero. Where we wish to detect a departure from a zero value of a predetermined size, this allows us to calculate the sample size necessary to do so with a given chance of success. In other words we can readily calculate the sample size necessary to obtain a "significant" result, say, 95% of the time, when the expected value of $\bar{z}$ is, in fact, not equal to zero.

Much controversy surrounds the use of significance tests (Barnett, 1975). Their main practical function is usually to help us make preliminary decisions, prior to studying the values of the estimates themselves. For example, in the above case the conventional "two-sided" significance test may be viewed as a test for the direction of the difference. In some circumstances it may well be reasonable to give first priority to determining the sample size with a view to giving a reliable decision about the direction of the difference. One of the difficulties with calculating sample sizes for significance tests, however, is the necessity for specifying the size of difference we want to detect. In some circumstances this difference may be determined from theoretical or practical policy considerations. Sometimes, previous evidence may be available or we may have to carry out a small pilot to obtain the required information. This does not arise when we only want to provide a confidence interval of a given length, although in both cases we need to have an estimate of the standard deviation. Assuming that significance tests may be regarded as preliminary steps in any analysis, we shall confine the following discussion to the precision of estimates of population values.

2.3.2 Specifying the Required Precision

If we leave aside sample design considerations, the obvious way of obtaining more precision or a narrower confidence interval is by increasing the sample size. If we know what precision we require, and have a good estimate of the standard deviation, then we can find the value of n, the sample size, which will give that precision. The precision we are seeking may be for a simple mean value, for the difference between two means, for the prediction of a child's or adult's height, for a percentile estimate, etc.

Methods of carrying out the calculations for the first three estimates mentioned can be found in many elementary textbooks on statistics and we shall only make general reference to them. We shall, however, go into more detail for percentile estimates.

To help determine the size of confidence interval to aim for, it is often useful to study the relative precision, obtained by dividing the confidence interval by an estimate of the mean value of the measurement. (The standard error divided by the mean value is often called the coefficient of variation of the mean.) For example, if the average height of a population of children of a given age is known to be about 120 cm and we have an estimate of the standard deviation in the population of 6·0 cm, the following table gives the 95% confidence interval width and relative width for different sizes of a simple random sample.

With a sample size of 500 we obtain a relative precision of about 0·4% whereas with a sample of only 10 the relative precision is about 2·5%. Such calculations can therefore give us a rough idea of the size of sample we might need, so long as we are prepared to commit ourselves to a level of relative precision. Similar calculations can be carried out for the difference of two means representing, for example, the change between two ages, the divisor here being the mean value of the difference. When predicting the value of adult height, say, we may divide the confidence intervals for predicted values by the mean predicted values.

Table 2.3 Sample Sizes for Specified 95% Confidence Interval Widths

	Sample Size				
	10	50	100	500	1000
Confidence interval width	3·04	1·36	0·96	0·43	0·30
Width relative to mean (%)	2·53	1·13	0·80	0·36	0·25

Population standard deviation = 6·0 cm Mean = 120·0 cm

With measurements which change over time an alternative method of arriving at an estimate of precision is to compare the confidence interval with the average change over a convenient time interval, say a year. Thus in the above example, if the average growth in height over a year is 5·0 cm, then with 500 in the sample the confidence interval is about 1·0 months of growth whereas with a sample of only 10 it is about 7·3 months of growth.

2.3.3 The Precision of Percentile Estimates

When constructing percentile standards for developmental measurements two aspects of sampling must be considered. The first, which we deal with in Section 2.4.2, concerns the manner in which the sample is spread across the whole age range. The second is concerned with the sample size needed to obtain a stated accuracy at any given age, and we deal with this now.

Where the distribution of a variable is known to be normal then we can use the estimated sample mean and standard deviation at the age concerned to give estimates of the percentiles. For example, a consistent estimate of the 97th percentile is

$$\bar{x} + 1{\cdot}88S$$

and the standard error of this estimate is approximately $1{\cdot}7S/\sqrt{n}$. For large samples this enables us to construct confidence intervals in the usual way for different values of n.

In practice, however, even for a variable so close to having a normal distribution as height, the assumption of a normal distribution ceases to provide good estimates of the extreme percentiles (Goldstein, 1972). The simplest and, for large samples, the safest alternative procedure is an ordering method whereby all the sample values are arranged in ascending order of magnitude and we choose the value (or interpolated value) below which the required percentage of the measurements lie. If the sample does, in fact, come from a normal population then the standard deviation of this estimate is approximately $2{\cdot}5S/\sqrt{n}$, so that it gives a confidence interval about 50% wider than those above. Table 2.4 sets out the range of population percentiles covered by a 95% confidence interval for the 97th percentile for each method.

Table 2.4 Percentile range covered on average by a 95% confidence interval about the estimated 97th percentile where the distribution is normal

Sample size	100	200	500	1000	2000
Simple Ranking method	91·6–99·1	93·5–98·7	95·2–98·2	95·8–97·9	96·2–97·7
Using sample mean and standard deviation	93·8–98·7	95·0–98·3	95·9–97·9	96·2–97·6	96·4–97·5

Whereas the sample mean and standard deviation give a considerable improvement for small sample sizes, for samples of 1000 or over either method is acceptable. The possibility of a large bias, which would occur if we used the mean and standard deviation for a population which does not have a normal distribution, might well outweigh the higher apparent accuracy.

We can improve the accuracy of the ordering method as follows. We plot the cumulative distribution of the points on a suitable scale (for example normal probability paper) and then "smooth" the curve in the region of the percentile to be estimated. Often this can be done adequately by eye. A detailed description of this method can be found in Goldstein (1972), and if we adopt it, a sample of about 500 should give acceptable accuracy for most purposes.

2.4 Sampling With Respect to Age

When designing a study covering a specified age range, we need to consider the manner in which measurements are allocated to ages within the range.

Many ways of spreading the sample over the age range are possible and we begin the discussion by assuming that measurements are made at a small number of discrete occasions. Later on we shall extend this to consider continuous sampling over the age range.

2.4.1 A Fixed Number of Occasions

A popular design for many developmental studies consists of measuring each individual child once a year on his or her birthday. This provides a fixed set of occasions a year apart which is convenient and ensures that "seasonal" variations are eliminated. Extra occasions are often inserted, for example at those times during development, such as at adolescence, when rapid changes are taking place. We shall pursue that point and the topic of seasonal variation below.

To simplify matters, consider just two occasions labelled 1 and 2. Suppose also that we have decided to measure the same number of individuals (n) at each occasion and that we are interested in making the most precise estimates possible of the mean at each occasion and the difference between the means. In general, the measurements at the two occasions will be related and we assume that the value of the measurement on the second occasion can be predicted from that on the first by the regression equation

$$x_2 = a + bx_1$$

where x_2 is the second occasion value and x_1 the first. The measurements need not be the same at each occasion although when we consider differences between occasion means, it is only possible to interpret such differences if a common measurement scale is used.

Suppose that m individuals are measured on both occasions, and $n - m$ are measured on one occasion only. We may use the m common measurements to estimate the values of a and b in the above equation in the usual way. Using these values we can then apply the equation to all those measurements taken on the first occasion to predict their mean value on the second. We thus have two separate estimates for the mean on the second occasion. One is obtained from those measurements made at the second occasion only; the other is obtained from those measured at the first occasion. We then find a weighted combination of these to give the most precise estimate of the overall mean at the second occasion. The reverse procedure can be adopted to obtain the most precise estimate of the mean at the first occasion. The most precise estimate of the difference between the means, where it is appropriate to calculate it, is then simply the difference of these two estimates. This procedure can readily be extended to several occasions with different numbers sampled at each occasion. This method, with certain simplifying assumptions, was described by Patterson (1950), and Gurney and Daly (1965), Jones (1975) and Woolson *et al.* (1978) describe more general procedures. These methods assume that the correlations between occasions and the variances within occasions are known. In practice they will often be calculated from the sample itself, and where the sample is large this will have little effect on the estimates. For small samples, however, and especially where the number of common individuals is small, it may be necessary to take this into account (see e.g. Jessen, 1942).

When all individuals are measured at both occasions, i.e. a pure longitudinal study, then the best estimates for each occasion are simply those which use the measurements made at that occasion. The same is true when there are no common units, since there is now insufficient information for relating the measurements on the two occasions.

Reverting to the two-occasion case we can illustrate the effect of different sampling strategies. The variance of the best estimate of the mean on each occasion is

$$V = \frac{s^2}{n}\frac{(1 - qr^2)}{(1 - q^2r^2)}$$

where q is the proportion not common on each occasion, s^2 is the variance within occasions, which is assumed to be the same for each occasion, and r is the correlation between occasions. If the variances differ, then a simple scale transformation can be used to create equality. If the same measuring instrument is used, for example to measure height, and the variances differ and we wish to estimate the difference between the means then such a scale transformation is inappropriate. Patterson (1950) gives appropriate modifications for unequal variances. When $q = 1$ or 0 we have $V = s^2/n$ which is simply the variance of the mean of n sample values. When $0 < q < 1$ then $(1 - qr^2)/(1 - q^2r^2) < 1$ so that having some common elements give us a more accurate estimate of the mean. In fact the variance is smallest when

$$q = \frac{1 - (1 - r^2)^{\frac{1}{2}}}{r^2}$$

The variance of the best estimate of change is $2s^2(1 - r)/n(1 - qr)$. If r is positive, as is often the case in physical growth studies, this variance is smallest when $q = 0$, that is all individuals are measured at both occasions. If r is not positive, then we obtain the smallest variance when there are no common elements. Table 2.5 shows the optimum values of q and corresponding precisions for estimating the occasion means and the difference of the means, for different values of r.

It is evident from this table that as the correlation between occasions increases, the gain from optimum allocation increases. The gain is much more striking for the estimate of the difference than for the means at each occasion. In particular, the last two columns show that when r exceeds about 0·5, the optimum mixture of common and different individuals for estimation of the occasion *means* becomes very inefficient for the estimation of the *difference* between occasion means. When $r = 0{\cdot}9$, the optimum mixture for the occasion means is to have 30% in common but this is only about 37% as efficient as the optimum for the difference, when all are common. On the other hand, when all are common the resulting estimates for the occasion means are about 72% as efficient as the optimum. Thus, if we wish to use a sample both to estimate occasion means and occasion mean differences, we will need to achieve a compromise in terms of efficiencies. The compromise will depend on the relative importance we attach to each estimate. When r is less than about 0·5 it would seem reasonable to choose all elements to be common. For physical growth measurements, however, the value of r for yearly measurements will often be as high as 0·7–0·9 and in this case we will therefore need to choose our priorities carefully. When the simple two-occasion situation is extended to several occasions with possibly unequal numbers, variances, etc., the optimum allocation procedures become more complicated, but the same general features still hold.

Table 2.5 Relative precision[a] of optimum values of q for different values of r

r	Optimum value of q	Relative precision (%)	Difference between occasion means Relative precision (%)	
			q = optimum value for occasion means	$q = 0$
0·1	0·50	100	106	111
0·3	0·51	102	121	143
0·5	0·54	107	166	200
0·7	0·58	117	193	333
0·9	0·70	139	370	1000

[a] Relative precision is the variance when $q = 1$ divided by the variance for the optimum value. It is therefore equal to the relative increase in sample size needed to maintain the same precision if separate individuals are used at each occasion as opposed to the optimum mixture of separate and common individuals.

The above discussion has been concerned solely with estimates of means and differences between means assuming known correlations between occasions. For many purposes, however, interest centres on estimating the regression and correlation coefficients between occasions (see Chapter 4). It seems that in this case a policy of retaining the same units at each occasion is usually the most efficient (Ian Plewis, personal communication). We have assumed implicitly so far that the relative cost of remeasuring an individual is the same as measuring a new individual. This is not necessarily true and ought to be taken into account when planning a study. In practice, however, it may be very difficult to calculate costs for a long time ahead and unless there are good reasons for supposing large cost differences between earlier and later measured individuals, the assumption of equal costs would seem reasonable. There may also be problems of differential bias, non-response, etc., when measuring individuals, which will also need consideration. We shall discuss these problems in later sections.

We have also assumed that the sample we are dealing with is a simple random sample from the population. Often in practice, however, the sample will be complex and involve several stages. In a 2-stage design, for example, we have the choice of retaining or changing individuals, as well as retaining or changing primary sampling units. There may be considerable cost savings involved in retaining the same set of primary units on a second occasion and changing some of the individuals within the primary sampling units. Although a detailed discussion of such situations is beyond the scope of the present chapter, a useful general strategy seems to be to keep all Primary Sampling units, replace some of the second stage units and retain the relevant later stage units. The interested reader is referred to the papers by Singh (1968) and Chakrabarty and Rana (1974) for further details.

In Chapter 4 we consider "growth curve models" where polynominal curves are fitted to longitudinal data on a sample of individuals. The aim is to describe growth in terms of a few simple parameters and to make comparisons between groups. A sampling problem which arises is that of how to space the occasions so as to provide most efficient estimates and comparisons. Morrison (1970) investigates this problem for a number of different assumptions about the form of the underlying curves and shows for a range of assumptions that equally spaced occasions are optimal or near optimal.

Machin (1975) considers the relative efficiency of growth curve models where the total number of individuals measured is kept constant, but the number of occasions and hence the number of individuals per occasion varies. The general finding is that efficiency is increased by reducing the number of occasions, that is by increasing the number of separate individuals in the study.

2.4.2 Sampling for Percentile Standards

We discussed earlier the sample sizes needed to achieve a given precision for estimates of the extreme percentiles of a distribution at a given occasion.

Typically, however, percentile standards are constructed for a wide age range, estimates usually being made at a large number of occasions within the range and these then joined up by smooth curves. The problem, therefore, is how such occasions are to be chosen, how many individuals should be sampled, and how the sample should be spread over the age range. The following discussion will assume that the same measurement is made at each occasion. It is particularly applicable, therefore, to physical growth studies and we shall use growth measurements as illustrations.

We start by assuming that percentiles are to be estimated with the same relative accuracy at all occasions or ages in the range. It follows that we should choose equal numbers at each occasion, which leaves us with the problem of how to space out the occasions. For any given percentile, say the 50th, we can estimate the value of the measurement at each occasion but no estimates of the values between occasions can be made. Thus between any two adjacent occasions, the expected value of the 50th percentile will have to be interpolated from the values at the occasions themselves. We might do this simply by joining the values at the occasions by a straight line, or if other occasions are available, by a smooth curve through a set of three or more occasions. Assuming that the true 50th percentile curve is smoothly increasing or decreasing between the two adjacent occasions, it is clear that the maximum difference between the true and interpolated values is simply the difference between the values at the two occasions.

As the criterion for spacing occasions, we shall specify that this maximum difference is the same at each part of the age range, and this leads us to space occasions so that the change in value of the 50th percentile is the same between all adjacent occasions. This is therefore equivalent to sampling occasions proportional to the rate of change of the 50th percentile value with age. Strictly speaking, this might lead to different sampling procedures if we choose different percentiles, but this is unlikely to be important in practice. Similar considerations apply to "velocity" standards (see Chapter 6) where we will be led to sample proportional to the rate of change of the 50th percentile. A problem occurs at the approach to adulthood where growth slows down and eventually ceases. It will then be necessary to include an adult sample of sufficient size to give the required precision. A further problem occurs when several measurements with different patterns of change are made using the same sample, and some compromise will have to be used to determine the sampling fractions.

Table 2.6 shows the relative sample sizes for yearly age groups based on the 50th percentile estimates of stature using British standards (Tanner *et al.*, 1966). Thus the highest frequency of sampling is in infancy where growth is fastest, followed by the adolescent period (see Fig. 6.1). The extra "adult" sample has been excluded.

Rather than concentrate a sample at specific ages, it is usually more practical to spread the sample along the whole age range. This would be the case, for example, when sampling children in school classes. It also enables us to eliminate any seasonal variations and to provide better 'smoothing' of percentiles. We then proceed by estimating the percentiles at the centre of narrow age intervals using a method which allows for the average change which takes place during the interval (Goldstein, 1972)*. The intervals are chosen as follows.

Table 2.6 Approximate Sampling Fraction for Cross-sectional Standards at each Year of Age from 0–19 years

Year of Age	Boys (%)	Girls (%)
0+	17	21
1+	8	9
2+	6	7
3+	5	7
4+	5	6
5+	5	5
6+	5	5
7+	5	5
8+	5	5
9+	5	5
10+	5	6
11+	5	6
12+	5	6
13+	6	4
14+	6	2
15+	4	1
16+	2	–
17+	1	–
Total	100	100

Ignoring the approach to adulthood, we choose a year when growth is slowest and select a large enough sample for that year to give precise estimates of the extreme percentiles. From Table 2.6 we might therefore select the age of 7+. If we act in accordance with the suggestion made in section 2.3.3, we

* Note that the expressions for estimates of β, α are incorrect on page 257 of Goldstein (1972). The correct expressions are:

$$\hat{\beta} = \sum_{i,j} w_i(t_i - \bar{t})(y_{ij} - \bar{y}) / \sum_{i,j} w_i(t_i - \bar{t})^2$$

$$\hat{\alpha} = \bar{y} - \hat{\beta}\bar{t}.$$

might take 500 children of each sex in this age group. To allow for seasonal effects estimation should be carried out over whole years so that a one-year age group is the smallest interval. Also, at this age, growth over one year is approximately linear, thus allowing a simple estimation procedure. The other age groups are then sampled accordingly. Where growth is fastest, namely under one year, the sample becomes relatively large. Since growth is changing very rapidly, a one-year interval will be too large and much narrower intervals should be used, for example one month. The seasonal effects here are likely to be very small relative to the change in growth itself and can probably be ignored.

In order to provide a smooth growth curve, we can form a moving average set of estimates by overlapping the age intervals. For example, we can choose the intervals 7·0–8·0, 7·25–8·25, 7·50–8·50, etc., estimating the percentiles at the centres of these intervals and subsequently joining up the resulting estimates 0·25 years apart.

The discussion so far has assumed that the measurements are continuous ones. For discrete measurements such as the stages of puberty, we wish to estimate the percentage of the population in each of the discrete stages at each age (see Chapter 6). The optimum method of sampling the age range is complicated, and depends on prior knowledge of the approximate values of the percentiles. From a practical point of view, however, it would seem adequate to sample uniformly over the age range during which transitions between stages are known to occur. It we use a log-linear model to provide "smooth" predictions of the percentages in each stage, then the total numbers of individuals needed in order to estimate the value of any given percentile will be similar to those required to estimate the same percentile by the method of smoothing described at the end of section 2.3.3. Since we wish to provide accurate estimates of the extreme percentiles, a sample size of about 500, which should give an acceptable accuracy for the 3rd percentile, may be regarded as a minimum number.

2.4.3 Seasonal Variations

In most longitudinal studies where attempts are made to measure individuals at predetermined ages, there will be some variations about the intended or target age. Where this variation is small and random, it will increase the variability of the estimates but not invalidate the procedures described above. If the variation about the target occasion becomes large, however, serious problems may arise. First, even where the distribution of ages about the target age is symmetrical, the variability of the estimates may be considerably increased if change is particularly rapid. More often than not, however, the distribution will be skewed, owing to most individuals being measured late, and there may also be an association between time of measurement and the average value of the measurement. Secondly, for physical growth it is known that this is more rapid at certain times of year (Marshall and Swan, 1971) so that differing time periods between occasions will lead to differing estimates of growth rate.

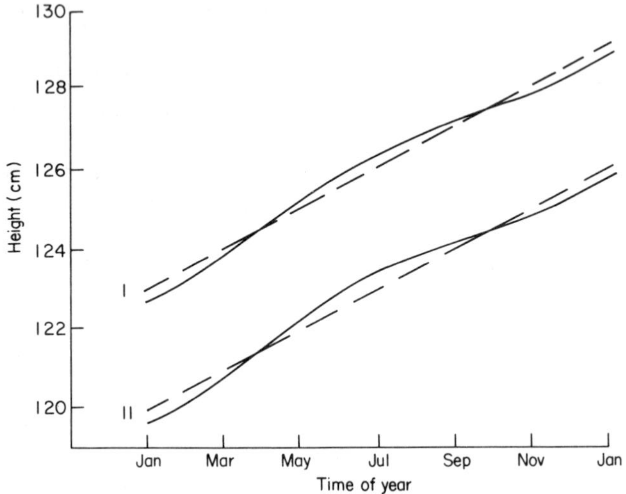

Fig. 2.2 Mean growth in height of two cohorts over one year. Cohort I is age 7·5 years and Cohort II is age 7·0 years in January.

Based on the data of Marshall and Swan (1971), Fig. 2.2 shows two curves of average growth in height for two "cohorts" aged exactly 7·0 and 7·5 years at the beginning of the year. Over the one year period it is assumed that they both increase by 6·0 cm so that there has been no secular trend operating over the half year separating the cohorts. We also see that the fastest rate of growth is in the spring. Thus an estimate of average growth rate based on a period of less than a whole year will give a biased estimate. For example, using either cohort, if we estimate a growth rate based on the six-month period from January to July, we obtain a value of 7·2 cm per year, which is 20% too large. Likewise, if we base an estimate on the difference between the older cohort in April and the younger in July we obtain a value of 5·4 cm per year, which is 11% too small.

We have similar problems when we wish to estimate the growth attained. For example, the average height of children who are 7·5 years in January (from the older cohort) is 122·7 cm whereas the average height of children who are 7·5 years in July (from the younger cohort) is 123·3 cm. This clearly has certain implications for the construction of standards for growth attained and, in particular, for the spacing of measurements over time. A similar phenomenon has been observed with tests of school attainment where some average attainment scores actually appear to decrease between the spring and the summer for 11-year-old children. (Goldstein and Fogelman, 1974.)

If we wish to estimate such seasonal effects it is clear that we need to measure individuals who are the same age at different times of the year from different cohorts, or alternatively individuals from the same cohort at different times of year. If we wish to avoid including any secular trends, the latter method is to be preferred. For periods of less than one year we will be concerned mainly with estimating single-occasion means rather than rates of change or correlations. We need not limit the sampling to specific occasions, but rather should spread sampling uniformly over the whole year in order to obtain more detailed estimates of the pattern. In addition, if we are also interested in change over a one-year period, then we can arrange for some or all of the individuals to be remeasured exactly one year later.

We have now arrived at a sampling scheme which is similar to one with yearly occasions, except that the occasions are now no longer at a specified age. Rather they are defined as any age within a year, subject to a uniform sample being spread over the year and measurements repeated at intervals of a year. In this way we largely solve both the problems outlined at the beginning of this section. If we regard the period between consecutive birthdays of a child as an occasion, we can readily apply the theory outlined previously to give optimum sampling schemes. An alternative would be to measure all individuals at the same time of year, selecting from different cohorts so that the age distribution covered exactly one year, for example 7·0 to 8·0 years. Finally, a combination of the two schemes could be used, sampling the same set of cohorts during the year. This approach seems to have been little tried in longitudinal studies, although it seems to possess advantages over the usual approach.

Before leaving this topic we mention briefly the problems which can arise from a failure to take account of seasonal effects in birth cohort studies. As an example we use the National Child Development Study, consisting of children born during the second week of March 1958. The children born at this time had a wide range of gestation lengths and hence a wide time span during which conception occurred. For example, a baby of short gestation of 34 completed weeks will have been conceived two months after a late gestation of 43 weeks, that is at the end of June 1957 as opposed to the end of April 1957. One consequence of this is that a change in conception rate between April and June 1957 will result in a biased estimate of the distribution of gestation length based on March 1958 births. If, as the evidence suggests, the conception rate was rising during this period, then the short gestations will tend to be slightly over-represented in the sample. Another consequence arises from the fact that specific periods of pregnancy are associated with the season of the year. For example, the fourth month of pregnancy occurs in August 1957 for the long gestations, but in October 1957 for the 34-week gestations and even later for the very short gestations. This early part of pregnancy is thought to be more vulnerable in terms of infections, etc., which may cause even minimal damage to the foetus, and since it is the short gestations which pass through the early part of pregnancy during the autumn when infections tend to be more prevalent, we might expect a resulting higher incidence of related difficulties and abnormalities in these births. Hence, an abnormality which is independent of gestation length *per se*, may in fact be associated with it through the association of conception time with season of the year. Without very detailed and accurate information about pregnancy infections (and other relevant data) we shall be unable to unravel these factors. These considerations, however, appear not to have caused serious problems in the National Child Development Study (Goldstein and Wedge, 1975).

2.5 Some Problems of Design and Inference

In this section we raise some of the issues of design and inference which are especially pertinent to observational studies of human populations. These issues arise from the typical inability of such studies randomly to assign subjects to "treatments" with the result that observed differences or associations between variables may not directly be interpreted as "causal" since they may be confounded with other related factors which may or may not have been taken into account. Much of the analysis of such studies is therefore concerned with making allowance for such factors in terms of measured variables, using techniques such as the analysis of covariance described in Chapter 4. The following discussion will centre on longitudinal designs based on relationships between measurements made on two occasions. No essentially new issues arise when relationships between more than two occasions are involved. Also, similar considerations apply to studies of growth curve type models, although in practice these tend not to be used in order to answer questions about causality.

2.5.1 Randomized Experiments

One of the simplest examples of a randomized experiment consists of two "treatments" or one "treatment" and a "control" with individuals from a given well-defined population being assigned at random to each "treatment". As a result of the experiment, measurements on a criterion variable are available for all individuals and the mean values for the two treatment groups are compared. The treatment might be part of a clinical trial to compare two drugs or, for example, part of an educational experiment to compare different methods of teaching children the rules of arithmetic. The statistical analysis of such experiments is essentially straightforward and well known, although practical complications do arise because of failure to meet distribution assumptions, etc. Nevertheless, the process of random assignment guarantees that statistical procedures, such as significance tests or the estimation of a confidence interval for a mean difference, allow one to make valid (probabilistic) inferences about the population from which the sample is drawn. Thus, a large difference in the means of measurements using two drugs with a narrow confidence interval, will allow an experimenter to infer that one of the drugs is superior to the other. Even in this case, however, we note that all experiments take place at a given historical time and, usually, in a circumscribed geographical area. The same experiment in other places and times might not yield the same results, and the general need to replicate experiments is widely recognised.

More elaborate randomized experiments can be designed and, in particular, they may involve repeated measurements and so entail methods of analysis appropriate to longitudinal designs. The parameters of such models can be estimated and compared as described in Chapter 4 and the same kinds of inferences made as in the simple case above. Let us consider the experiment mentioned above to compare two methods of teaching the rules of arthmetic.

Suppose that the experiment is designed to last for a year, and that two samples each consisting of 250 seven-year-old children, are selected randomly from the schools of an education authority. Two suitable tests of knowledge of the rules of arithmetic are devised and one is administered to each child at the start of the experiment and the other a year later at the end. (We ignore for now the technical problems which might arise such as that of "collusion" between subjects during the experiment.) A valid comparison between the teaching methods would be to compare, for example, the mean scores for the two methods at the end of the year. The initial test (or pre-test) scores of the children, however, will not be equal, even though the random allocation guarantees that no biases are present. By taking into account each individual's pre-existing ability we should be able to account for some of the variability in the final test (or post-test) scores and thus increase the sensitivity and accuracy of the comparison. The analysis of covariance was introduced by Fisher (1932) for this purpose, and the simplest model can be written, using the same notation as in Chapter 4,

$$y_{ij} = \alpha_i + \beta x_{ij} + \varepsilon_{ij} \tag{2.1}$$

where the subscript i refers to the groups being compared and j to the individuals within these groups.

If model (2.1) fits the data, then for two groups, because of the random initial assignment, the difference $\hat{\alpha}_1 - \hat{\alpha}_2$ is an unbiased estimate of the difference between the post-test means, say $\mu_1 - \mu_2$. As the correlation between the pre-test and post-test increases, so the standard error of this difference decreases. When the assignment is not random then $\hat{\alpha}_1 - \hat{\alpha}_2$ is no longer an unbiased estimate of the difference $\mu_1 - \mu_2$ unless $\beta = 0$. In the next section we look at designs which are not based on random assignment.

2.5.2 Quasi-experimental and Observational Studies

In many situations, especially when dealing with human subjects, it is either impossible or undesirable to allocate subjects randomly to treatments. For example, if we wish to compare the effects of different styles of teaching on the attainments of children, we may have to accept an allocation of classes to different styles which is determined by teachers' availability, administrative convenience, etc. It may be, for example, that a particular policy has resulted in teachers with certain styles being allocated to classes containing high ability children, so that a random or near random assignment cannot be assumed. In this situation we need to consider very carefully the interpretations which can be made.

Campbell and Stanley (1963) analyse and classify the different kinds of processes which are involved in interpreting such situations. They use the term "quasi-experiment" to imply that the aim of a design is that of a truly randomized experiment but without the simplicity of interpretation which full randomization allows. Much of this discussion is not peculiar to longitudinal studies, and some of the points are also discussed elsewhere in this book. For our present purposes the important problem is that which arises through the possible confounding between pre-treatment variables and the treatment given.

If we knew that the treatment groups differed on only one pre-treatment variable which was related to the post-treatment measurement, for example reading attainment, then we could make due adjustment for this in the subsequent analysis. If this variable were the reading attainment prior to treatment and if the relationships with post-treatment reading score were linear and parallel, then we have the model (2.1). Since for each given first-occasion reading score the treatment groups do not differ on any variable related to the occasion 2 score, we could interpret differences between the treatments as being *results* of the treatment. We see, therefore, that the statistical analysis is identical with the random assignment case, but the logic behind the equations (2.1) is now different. In particular, it relies upon the assumption that the chosen pre-treatment measure is the only one which it is relevant to consider. In practice we very rarely have such certain knowledge and we are faced with the problem, common to most observational studies of human populations, of disentangling interrelated variables in an attempt to isolate causal connections. Whether a "quasi-experiment" is viewed as approximating to a properly randomized experiment or whether any causal connection is thoroughly obscured by confounding variables will depend on the knowledge and circumstances attaching to the particular situation. Some examples are given by Campbell (1963) and a useful general discussion of the problems of adjustment in non-randomized studies is given by Cochran (1965). Rubin (1977) considers the intermediate case of partial randomization, where random allocation to treatment takes place within a set of selected values of a covariate, or combination of covariates. His model also allows that all the individuals at a selected value are assigned to just one treatment. In order to carry out this analysis it is necessary to assume that the form of the relationship is known over the whole relevant range of the covariate, so that in practice the selected values will have to be carefully chosen for each treatment to ensure this. Where the assumptions of the model are all satisfied, then we can make similar interpretations as in the fully randomized experiment.

In summary, therefore, once we leave the area of fully randomized experiments, considerable care is required before quasi-experiments can be interpreted as if they were randomized experiments, and this brings us to the third situation, where we are not attempting to carry out any kind of "experiment", but rather to estimate relationships over time.

In this kind of study the basic statistical models are similar to those we discussed above. Sometimes the purpose differs, for example when we wish to predict a second occasion measurement for a given first occasion measurement, as when constructing conditional developmental standards (see Chapter 6). Alternatively, we may be interested in making "causal" inferences, as in the quasi-experimental case by making suitable adjustments for first occasion measures. The important difference is that the observational longitudinal study has no control at all over which treatments are assigned to which groups. By contrast, although in a quasi-experiment a completely random assignment is not possible, some control of treatment assignment is typically present, and also there is a definite point, usually immediately following the first occasion measurement, when treatments are commenced. In an observational study, the two occasions do not normally coincide with the start and finish of a treatment. In Chapter 4 we give an example of an observational longitudinal study and discuss the kinds of interpretations which can be made.

2.6 Non-response Bias

Many stages intervene between the design of a study and the analysis of the data, and at each one errors may occur. Those which occur at the time of sampling and measuring tend to be often the most important and have the greatest potential for invalidating the results of analyses.

A non-respondent is a selected individual for whom data is unavailable. We can distinguish three main situations which can give rise to this. First, it may be impossible to trace the individual, for example he has moved. Secondly, the individual may refuse to cooperate, and thirdly the individual may become excluded from the study, for example because of death.

If the individuals who fail to provide information were a random subset of the sample, no bias would result although the design would become modified (see, for example, Lehnen and Koch, 1974). Unfortunately, in nearly all surveys the non-responders have different characteristics from the remainder. If we knew how the non-responders differed we could use the information from the remainder of the sample to make appropriate corrections to our estimates. For example, if we knew that the average age of the non-responders was higher than the remainder, and that this was the sole factor responsible for the non-response, then we could weight up the remainder of the sample to correct for the bias in those variables related to age.

Most procedures for dealing with non-response bias are attempts to discover which characteristics of the non-responders are different from the remainder, to assess the likely effect of these differences and, where appropriate, to make suitable corrections. For example, if repeated call-backs are used to obtain measurements, we will often find that the more difficult the subject is to measure, the more different are his characteristics. By extrapolating the relationship between difficulty to measure and the values of various characteristics, it may be possible to obtain a good estimate of the values of the characteristics of those who do not respond at all. (See e.g. Houseman, 1953.)

With a longitudinal study we are, in some respects, in a worse position regarding non-response than with a single cross-sectional study, since non-response bias can occur at every occasion. As the study progresses the cumulative effect of non-response may lead to cumulative bias. This will affect both the estimate of cross-sectional parameters and also the estimates of change and relationships across time. Williams (1970), for example, quotes evidence from a longitudinal study showing that over time there is a tendency for large families to drop out of the study. Williams and Mallows (1970) suggest that for measurements which are highly associated with response, even an overall response rate as high as 90%, which is often regarded as satisfactory, can lead to very biased estimates of change. Unfortunately, it is often just those groups of the population who show the most change over time who are also the most mobile, better educated, etc., and more likely to be untraced, unavailable or refuse to cooperate.

It is important to distinguish between these three groups of non-responders. There is evidence (see e.g. Labouvie *et al.*, 1974) that children whose parents refuse to cooperate on the second or subsequent occasions have school attainments at least as high as those with information, whereas those who are untraced and have no available data do worse. This is supported by evidence from the National Child Development Study (Goldstein, 1976) which shows that those children who refused to cooperate in the study at the age of sixteen had reading, mathematics and IQ scores at 11 years, four to six months in advance of those who did cooperate. Those who had been traced but had supplied no available data at 16 years were four to six months behind and those who were untraced at 16 years were nearly a year behind at 11 years. In terms of height measured at 11 years there were also some detectable differences between these groups. These were, however, relatively small and, although it would be rash to generalize from this study alone, it appears as if physical growth measurements might be less affected by non-response than mental measurements. There was also an indication in this study that the refusers had a slightly faster rate of mental development before 11 years and those with no data a slightly slower rate of development, suggesting that similar differences might hold for the period from 11 to 16 years.

We see here how internal evidence using the longitudinal elements of a study may be used to study the effects of different response patterns. By itself it can say little about initial bias, nor can it give a complete picture of change bias. By sampling sufficient new individuals at each occasion, however, we can obtain further information about cumulative occasion bias and also investigate the bias in estimates of change. In any longitudinal study where non-response is likely to occur, it will be worthwhile to sample extra new individuals for this purpose. In some circumstances there may also be data from other sources which can be used to check for bias. Sobol (1959) uses this method to show that, after four occasions each six months apart, the attitudes of an adult sample to economic affairs were little different to those of comparable but newly selected individuals.

In a quasi-experimental study, because the groups are undergoing different experiences, this can often give rise to *differential* response rates and great care has to be taken to minimise this by careful tracing, call-backs, etc. Using the initial pre-treatment measurements to study response bias will generally be less satisfactory here since the point of the study itself is to bring about large changes.

2.7 Repeated Response and Context Bias

The repeated measurement of a sample of individuals may alter the natural history of their growth and development. This may occur in a physical growth study, for example, because a set of detailed measurements may uncover growth irregularities or morbidity patterns which will receive treatment they would not otherwise have had. The increased awareness which usually results from taking part in a study may also affect the measurements being studied. Bailar (1975) suggests that the length of exposure to a study affects both cross-sectional and longitudinal estimates of employment. With mental measurements, such factors as memory of previous responses and practice effects will influence subsequent measurements. It is therefore worth considering setting aside portions of the sample measured on the first occasion, to be remeasured once only at later occasions in order to study the effect of repeated measurements on the estimates. Such a procedure was also mentioned in Chapter 1 in connection with piloting.

The context in which a measurement is made may influence the value obtained. For example, in a questionnaire interview the order in which questions appear can markedly influence the response. Where a question is repeated on successive occasions to obtain measures of change, we need to take care that the fact that it may be appearing alongside new questions does not of itself cause a change in response. Careful anticipation of this problem at the various piloting stages is advisable. Schuman (1974) gives a detailed discussion of this problem in attitude surveys.

The historical context of a measurement is also very important for some kinds of measurements. Tests of reading attainment in children, for example, may produce varying responses over time both because the use of language itself changes, and changes take place in school curricula. A discussion of some of these difficulties can be found in Start and Wells (1972).

2.8 Secondary Sampling

Whereas most of this chapter has been concerned with overall sampling strategies, it will very often be necessary in a longitudinal study to defer final decisions regarding sampling on future occasions until, say, some results accumulate or the response rate becomes known. Often not very much can be said about such possibilities except that there should be an awareness of them from the outset. With large studies of the cohort type where large amounts of data are collected at relatively infrequent intervals, there is often a provision for special groups to be subsampled and studied more intensively.

The decision to do this, however, can usually not be made at the outset. It may have to wait, for example, until there is relevant information to define subgroups, or there may arise new issues of theoretical or practical interest which can help to determine the course of the study. Conversely, we may often be justified in including measurements at the start of the study whose sole function is to provide basic data for possible later use. We might, for example, collect basic information concerning a study child's siblings to allow for a later study which would follow them up. Also we might collect data which had no immediate relevance to the study but which might become relevant later. Especially with long term studies, it is often the case that the interests of researchers and policy makers will change during the course of the study. Some allowance for this seems desirable provided, of course, that it does not result in a massive and uncritical collection of information.

References

Bailar, B. A. (1975). The effects of rotation group bias in estimates from panel surveys, *Journal of the American Statistical Association*, **70**, 23–30.

Baltes, P. B. (1968). Longitudinal and cross-sectional sequences in the study of age and generation effects, *Human Development*, **11**, 145–171.

Barnett, V. (1975). "Comparative Statistical Inference", Wiley, London.

Buss, A. R. (1973). An extension of developmental models that separate ontogenetic changes and cohort differences, *Psychological Bulletin*, **80**, 466–479.

Campbell, D. T. and Stanley, J. C. (1963). "Experimental and Quasi-experimental Designs" Rand McNally & Co. Chicago.

Chakrabaty, R. P. and Rana, D. S. (1974). Multi-stage sampling with partial replacement of the sample on successive occasions, *Proceedings of the Social Statistics Section of the American Statistical Association*, 262-268.

Cochran, W. G. (1965). The planning of observational studies of human populations, *Journal of the Royal Statistical Society*, A **128**, 234–266.

Davie, R., Butler, N. R. and Goldstein, H. (1972). "From Birth to Seven", Longman, London.

Douglas, J. W. B. (1964). "The Home and the School", McGibbon and Kee, London.

Douglas, J. W. B. and Blomfield, J. M. (1958). "Children Under Five", George Allen and Unwin, London.

Fisher, R. A. (1932). "Statistical Methods for Research Workers", Fourth Edition, Oliver & Boyd, Edinburgh.

Goldstein, H. (1972). The construction of standards for measurements subject to growth, *Human Biology*, **44**, 255–261.

Goldstein, H. (1976). A study of the response rates of sixteen-year-olds in the National Child Development Study, *In* "Britain's Sixteen-year Olds" (Fogelman, K., ed) National Children's Bureau, London.

Goldstein, H. (1979). Age, period and cohort effects: A confounded confusion, *BIAS*, **6**, 19–24.

Goldstein, H. and Fogelman, K. R. (1974). Age standardisation and seasonal effects in mental testing, *British Journal of Educational Psychology*, **44**, 109–115.

Goldstein, H. and Wedge, P. J. (1975). "The British National Child Development Study", World Health Statistics Report, **28**, 202–212.

Gurney, M. and Daly, J. F. (1965). A multivariate approach to estimation in periodic sample surveys, *Proceedings of the Social Statistics Section of the American Statistical Association*, 202–257.

Holt, D., Richardson, S. C. and Mitchell, P. W. (1978). The analysis of correlations in complex survey data; unpublished mimeo, Dept. of Social Statistics, University of Southampton.

Houseman, E. E. (1953). Statistical Treatment of the non-response problem, *Agricultural Economics Research*, **5**, 12–18.

Jessen, R. J. (1942). Statistical investigation of a sample survey for obtaining farm facts, *Iowa Agricultural Experimental Station Research Bulletin*, 304.

Jones, R. G. (1975). "Time Series Analysis of Repeated Surveys", MSc Thesis, University of Southampton.

Jordan, J., Ruben, M., Hernandez, J., Bebelagua, A., Tanner, J. M. and Goldstein, H. (1975). The 1972 Cuban National Child Growth Study as an example of population health monitoring: design and methods, *Annals of Human Biology*, **2**, 153–171.

Kish, L. (1965). "Survey Sampling", Wiley, New York.

Kish, L. and Frankel, M. R. (1974). Inference from complex samples, *Journal of the Royal Statistical Society*, B **36**, 1–37.

Labouvie, E. W., Bartsh, T. W. Nesselroode, J. R. and Baltes, P. B. (1974). On the internal and external validity of simple longitudinal designs, *Child Development*, **45**, 282–290.

Lehnen, R. G. and Koch, G. G. (1974). Analyzing panel data with uncontrolled attrition, *Public Opinion Quarterly*, **38**, 40–56.

Machin, D. (1975). On a design problem in growth studies, *Biometrics*, **31**, 749–753.

Malina, R. M., Hamill, P. V. B. and Lemenshaw, S. (1974). Body dimensions and proportions, white and negro children, 6–11 years, *Vital and Health Statistics, Series* 11, US Dept of Health Education and Welfare.

Marshall, W. A, and Swan, A. V. (1971). Seasonal variation in growth rates of normal and blind children, *Human Biology*, **43**, 502–516.

Mason, K. O., Mason, W. M., Wimborough, H. H. and Poole, W. K. (1973). Some methodological issues in cohort analysis of archival data, *American Sociological Review*, **38**, 242–258.

Morrison, D. F. (1970). The optimal spacing of repeated measurements, *Biometrics*, **26**, 281–290.

Patterson, H. D. (1950). Sampling on successive occasions with partial replacement of units, *Journal of the Royal Statistical Society*, B **12–13**, 241–255.

Rubin, D. B. (1977). Assignment to treatment groups on the basis of a covariate, *Journal of Educational Statistics*, **2**, 1–26.

Schaie, K. W. (1965). A general model for the study of developmental problems, *Psychological Bulletin*, **64**, 92–107.

Schlesselman, J. J. (1973). Planning a longitudinal study, sample size determination, frequency of measurement and study duration, *Journal of Chronic Diseases*, **26**, 553–570.

Schuman, H. (1974). Old wine in new bottles: some sources of response error in the use of attitude surveys to study social change, Paper prepared for SSRC Research Seminar Group Meeting at University of Surrey, 4 April, 1974.

Singh, D. (1968). Estimates in successive sampling using a multistage design, *Journal of the American Statistical Association*, **63**, 99–112.

Sobol, M. G. (1959). Panel mortality and panel bias, *Journal of the American Statistical Association,* **54**, 52–68.

Start, K. B. and Wells, B. K. (1972). "The Trend of Reading Standards", NFER, Slough.

Tanner, J. M., Whitehouse, R. H. and Takaishi, M. (1966). Standards from birth to maturity for height, weight, height velocity and weight velocity: British children, 1965, *Archives of Disease in Childhood*, **41**, 454–471, 613–635.

Wall, W. D. and Williams, H. L. (1970). "Longitudinal Studies and the Social Sciences", Heinemann, London.

Williams, W. H. (1970). The systematic bias effects of incomplete response in rotation samples, *Public Opinion Quarterly*, **32**, 593–602.

Williams, H. W. and Mallows, C. L. (1970). Systematic biases in panel surveys due to differential non-response, *Journal of the American Statistical Association,* **65**, 1338–1349.

Woolson, R. F., Leeper, J. D. and Clarke, W. R. (1978). Analysis of incomplete data from longitudinal studies, *Journal of the Royal Statistical Society*, A **141**, 242–252.

3

The Choice of Measurements Over Time

This chapter deals with the special problems which arise when a measuring instrument is used on more than one occasion. By "measuring instrument", we mean any method of collecting information, whether, for example, by use of a stadiometer to measure height, a set of items to measure reading attainment, or a question about political attitudes. The problems arise in four ways.

Firstly, an instrument which is appropriate at one age or time may not be appropriate for measuring the same underlying characteristic at another age or time. This can be a serious difficulty, for example, with measurements of mental and behavioural characteristics, where a test or questionnaire may be designed for a restricted age range. Outside this age range some or all of the items comprising, say, a reading test may be inappropriate because they are too easy or too difficult, so tending to elicit the same responses from every member of the population.

Secondly, a reasonably high level of participation from the subjects may be required in order to ensure comparable results. When a subject is questioned, for example, the understanding of the question and hence the interpretation, may change systematically with age and over time, as well as across different subgroups of the population. With some physical measurements also, such as measures of strength, a certain amount of motivation is required of the subject.

Thirdly, there is the problem of measurer error. Measurers will tend, on average, to differ systematically from one another in the responses they obtain from the same set of individual subjects. Subjective judgements of distance in physical measurement, or differences of presentation in interviews, may be related to the response (Hoinville and Jowell, 1975). Careful training and detailed instruction will help to reduce such differences, and in a carefully designed study it may be possible to allocate measurers among subjects so as to avoid confounding measurer differences with group differences. The question of quality control in longitudinal studies is discussed in Chapter 1.

Fourthly, many responses to questions, or assessments of physical features, for example, will be coded or classified into one of a number of categories by a researcher, prior to analysis. Often a subjective judgement will be necessary to allocate "boundary" cases, and this may mean that different coders will produce different distributions of coded responses from the same set of basic data. This problem is also discussed in Chapter 1.

The remainder of this chapter discusses the first two of the above problems and their implications for the designer of measuring instruments. We shall not, however, describe the detailed construction or usefulness of particular instruments, and the reader is referred to specialised descriptions for particular fields of interest. (See for example, Thorndike (1971) for mental measurements; Moser and Kalton (1971) for social survey measurements; and Cameron (1978) for anthropometric measurements.)

3.1 Types of Measurement Scales

The term "measurement" can have different meanings. In popular usage it is often restricted to what are technically known as "interval" and "ratio" scales, or even just to continuously varying interval and ratio scales such as those of temperature and height. We shall use the term in a wider sense, and we distinguish four types of measurement scale as follows;

(i) Nominal Scale: This includes measurements such as sex and geographic region, which are simply classifications of a population into mutually exclusive categories which have no natural or defined ordering.
(ii) Ordinal Scale: This includes measurements in ordered categories such as questions on attitudes, where a subject is asked to say, for example, whether they "agree with", "are indifferent to", or "disagree with" a statement. It also includes measurements such as that of bone development, where a number of stages of calcification can be distinguished. In both these instances there is a natural or defined ordering of the categories, but no obvious way of assigning values to the distances between them.
(iii) Interval Scale: This includes measurements such as IQ, where a numerical value is assigned to the measurement and where the *differences* between these values have a direct interpretation and may be used to compare individuals. Thus a difference of 10 IQ points may be compared with a difference of 20 IQ points.
(iv) Ratio Scale: This includes measurements which are made on an interval scale and which also have a "natural origin". Height and family size fall into this group, since they are both interval measurements and both have well defined zero points. This kind of scale allows us to attach significance to *ratios* of scale values.

A further classification of measurements into discrete and continuous is also made. The former includes nominal and ordinal scale measurements, and others, such as family size, which are measured in integer units. Continuous measurements are those interval and ratio scale measurements, such as height, with continuously varying values.

In general, the ways in which these types of measurement are analysed and interpreted will differ. The distinction between them, however, is not always very rigid, so that, for example, a measurement of one type can often be converted by a simple operation into a measurement of another type. Thus social class, which is usually measured on an ordinal scale, is sometimes assigned an interval scale by allocating the integer numbers, 1, 2,... to the classes in order. Alternatively, we could choose to convert a ratio scaled measurement, such as height, into an ordinally scaled measurement, by categorizing individuals into three groups; short, medium and tall. A more detailed account of measurement scales is given by Torgerson (1958).

3.2 Measurement Standardization and Transformation

As we move from measurements made on nominal or ordinal scales to those made on interval or ratio scales, we obtain measurements which have increasing utility in terms of the amount of information they convey. Thus, for example, we obtain more information from a measurement of height in centimetres, than from a classification into short, medium or tall. A considerable part of scaling theory (see Torgerson, 1958) is, in fact, devoted to the problem of transforming nominal or ordinal scale measurements to continuous interval scale measurements, and we shall discuss a particular example of transforming a set of ordinal scale measurements to an interval scale in a later section which deals with maturity scales.

With many interval scaled measurements there is an arbitrariness about the units in which the scale is expressed. This is illustrated by many mental measurements, where the "raw" test score has traditionally been transformed into a standardized score with certain properties, typically to have a population distribution with mean equal to 100 and standard deviation equal to 15. That particular transformation is an example of a "non-linear" transformation, by which we mean the following. First, a linear transformation of a variable or measurement x into a variable y is one which has the form,

$$y = a + bx$$

where, in the equation relating y to x, a and b are constants. A non-linear transformation is an equation relating y and x using other functions of x. For example,

$$y = a + bx^3$$

or

$$y = a + b \log x$$

are both non-linear transformations of x into y. They are also what are called "monotone" transformations; that is they preserve the ordering of the scale values, so that if $x_1 < x_2$ then $y_1 < y_2$.

Such transformations, however, do not preserve the ordering of *differences* between scale values. Thus, if $(x_2 - x_1) < (x_4 - x_3)$ it does not follow necessarily that $(y_2 - y_1) < (y_4 - y_3)$. If there is no particular reason for preferring one scale of measurement over another, which is a non-linear transformation of it, we can have quite different interpretations of data according to which scale is used. Special attention, therefore, needs to be paid to the real-life meaning which can be attached to particular scales. This is the case if we wish to work with interval scales. If we confine ourselves to analysing the order relationships of the measurements, then the same problem does not arise.

To illustrate this point, Table 3.1 gives perinatal mortality rate comparisons for two countries, based on data extracted from the World Health Organisation Health Statistics Annual (WHO, 1975).

Table 3.1 Perinatal deaths per 1000 total births

Country	Year 1954	1965
Austria	41·5	29·5
England and Wales	38·0	26·9
Difference	3·5	2·6

By subtraction, we see that the difference between the mortality rates for the two countries was greater in 1954 than in 1965. Suppose, now, that we apply the transformation,

$$y = \log_{10} x$$

where x is the perinatal mortality rate. We obtain the new transformed rates given in Table 3.2.

We now see that, on this scale, the difference between the countries was greater in 1965 than in 1954, a reversal of the previous difference. In this example it is not at all clear which scale is the more appropriate one for making statements about the changes in mortality rates. The differences on the log scale are, in fact, the logarithms of the ratios of the rates, so that a measure of relative risk is being used. The differences in mortality rates relate to differences in numbers of deaths. Any choice between measures should depend on the nature of the factors which are known to influence perinatal mortality, and on the nature of the relationship between such factors and the mortality rate, as well as on the purpose of the measurement itself.

Table 3.2 $\log_{10}$ (perinatal deaths per 1000 total births)

Country	Year 1954	1965
Austria	1·618	1·470
England and Wales	1·580	1·430
Difference	0·038	0·040

In other words, the scale of measurement adopted should, among other things, depend on our scientific knowledge, or our model, of the properties we are measuring. Where the appropriate knowledge is lacking, or at least is rather vague, then the scale adopted may contain a high degree of arbitrariness, and it will be important to recognise that any interpretation may be closely related to choice of scale. It would seem that many mental measurements have this problem, the purpose of the non-linear transformation of the raw scores being to cause the transformed scale to possess certain statistical properties, rather than conform to any theoretical model.

If, in the above example, we had used mean weights of children in two social classes at two different ages, we see how statements about changes in growth could depend on choice of scale also.

It should be pointed out that we are not now discussing the *prediction* of a variable, for example weight at one age, from the value of another variable, for example weight at an earlier age. In such a case, the purpose of any scale transformation is to provide a more efficient prediction. Thus if the predicted logarithm of the weight, after transforming back to the scale of weight, has a smaller variability than the predicted untransformed weight, then we would do better to use a logarithmic transformation.

If we accept the arbitrariness of a measurement scale are there any other criteria which might help in selecting one particular scale? Statistical distributional considerations such as normality and stability of variance may play some part, but the most useful criterion is to have a scale which leads to simple mathematical relationships with other variables. In regression analysis, for example, a transformation of the dependent variable is often used to eliminate "interactions", and so to provide a simpler relationship between the dependent and independent variables. Where the relationships are longitudinal ones, between different ages for example, the existence of a simple relationship can aid our understanding. Thus, for example, in the analysis of mental measurements it might be better in some circumstances to use the raw scores rather than the transformed scores. We shall have more to say on this in Chapter 5.

We should remember, however, that while simplicity is a useful criterion it is not sufficient, and the possibility of alternative interpretations based on different theoretical assumptions should be given emphasis. In many cases the interpretation of an analysis will be similar for a wide range of scales, so that the choice of a particular scale is not crucial. It is a useful technique to compare different scale transformations to see, in fact, whether this is true.

3.3 Age-specific Measurements

With certain measurements, such as height or skeletal maturity, the same instrument can be used at all ages with little ambiguity as to its relevance, thus allowing us to analyse and describe the relationship of the measurement to age. With other measurements, such as attitude scales or educational attainment tests, different measuring instruments will usually be used for different parts of the age range of interest and this poses some conceptual problems.

It is sometimes assumed that different measurements at each occasion are, in fact, measuring the same underlying attribute. In practice this may be a very questionable assumption and its validity will depend on the amount of theoretical support available. Even where the same instrument is used on each occasion, the interpretations of it may differ. A test of mathematical attainment, for example, may be used with children of different ages, and typically we will find that the younger children tend to answer only the easier questions whereas the older children can tackle the more difficult ones. Effectively, in this case, we would be using somewhat different measuring instruments, and we need to justify carefully any claim that the same underlying attribute is being measured.

It is, however, not always strictly necessary to obtain firm theoretical support for the use of different measures to assess the same quantity. It may be possible to argue, on pragmatic grounds, that the use of two particular measuring instruments on two different occasions, and the study of changes between them, give useful predictions or discriminate well between different groups of children.

A further justification for treating measurements as reflecting the same attribute occurs if the measurements themselves are thought of, and used, in assessing individuals *as if* they were reflections of the same attribute. Such a situation may occur in a school which conceives of reading attainment as an underlying attribute which changes in value, and which it is appropriate to estimate using different instruments. In this case, educational action may be based on the assumption of such a common underlying attribute, which may or may not have theoretical support, but which nevertheless is in practical use.

It is possible, also, to consider models based on the assumption of an unobservable, but underlying attribute or trait which exhibits a relationship with age, for example by a logistic curve of the type discussed in Chapter 4. By further specifying how "observed" measurements are functionally related to this underlying trait, we can define a "latent trait" model, where parameters can be estimated using a sample of individuals. An example of this approach is given by Bock (1976). We shall discuss models of this kind in Chapter 5, but it is worth remarking here that there are difficult problems in finding sensible interpretations of any results from such models, for example, when we move away from very narrowly defined psychological attributes to educational attainments (Goldstein and Blinkhorn, 1977).

If we are satisfied that a measurement is, in an appropriate way, reflecting the same quantity on each occasion, we may treat it broadly in one of two ways. Firstly, we may set up models which relate measurements across occasions, and such models are discussed in Chapter 5 so that we need not discuss them further here. Secondly, we may wish to form a common scale applicable to all occasions so that the models in Chapter 4 can be applied. To illustrate this procedure, suppose that there are three separate tests, all assumed to measure the same underlying component of mathematical ability, and which are considered appropriate over the following age ranges.

	Age Range (yr)
Test A	5·0 – 9·0
Test B	8·0–12·0
Test C	11·0–15·0

We would like to be able to express all these measurements on a common scale. A condition for this to be possible is, that in the age range where the tests overlap we should, in principle, be able to match every score on test A with a score on test B and every score on test B with a score on test C. This means that we should be able to find a monotone transformation of scale A to scale B and scale B to scale C. Also, the original scales should possess the same transformation from one to the other over the whole age range of overlap. Thus, for example, if a particular score on test A is equivalent to a particular score on test B at age 8·0, the two scores should also be equivalent at all ages from 8·0 to 9·0 years.

There are two different general solutions to the problem of a common scale. One solution is to provide separate age standardizations for each test, so that the standardized measurement has an identical population distribution at each age. This is typically done in the case of mental measurements, which are transformed to have a normal distribution with a constant mean and standard deviation. More generally, where we have available population standards for the age range (see Chapter 6), we can assign a percentile value to any measurement, and then convert this into, say, the corresponding value for a normal distribution. Thus, for example, a boy at the 75th percentile of the standards would be assigned an equivalent value of 110·1 from the normal distribution with a mean equal to 100 and a standard deviation equal to 15. We are left, therefore, with a variable or transformed measurement which has a mean and a standard deviation which do not change with age, and we may study the changes in, say, mean values of population subgroups over time.

The second solution to the problem involves making use of the information provided by the overlap of age ranges in order to provide a common scale. Consider the age range 8·0 to 9·0 where the above tests A and B overlap. We will assume that, apart from sampling fluctuations and other small random errors, we can satisfy the scale transformation condition given above. We might estimate the relationship, for example, by giving both tests to a large sample of eight-year-olds and then analysing the relationship between the resulting scores. By further dividing this age range into narrower age ranges, and comparing the estimated relationships between ranges, we would also be able to provide a verification of the above assumption, as well as estimating the overall relationship between the scores, so that a score on one scale can be converted or transformed into a score on the other. Having converted scores on test B to the scale of test A, say, we can do likewise for test C to the scale of test B, and hence to the scale of test A, which becomes the common scale. Note that we do not necessarily require a linear relationship to exist, although the existence of linear transformations between scales will generally simplify any interpretations.

There is, however, a drawback which may make this method difficult to use in practice. It is possible that, in the age range of overlap, only a part of the possible range of scores will be present, the upper part of the range for test A and the lower part of the range for test B. In such a situation we can extrapolate, to the remainder of the ranges, the relationship found over the restricted score ranges. Whether this is permissible or not will have to be decided separately for each situation. Figure 3.1 illustrates such an extrapolation.

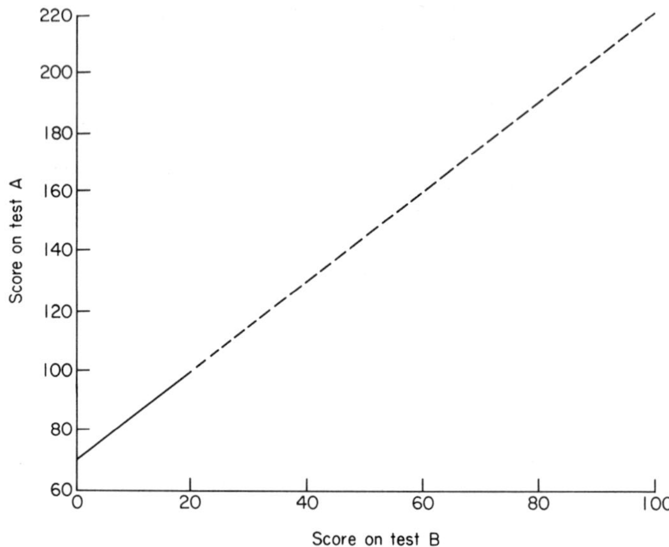

Fig. 3.1 Straight line $y = 70 + 1{\cdot}5x$ fitted to two hypothetical tests, A and B, with scores y and x for the range $0 \leqslant x \leqslant 20$ (continuous line), and extrapolated to the range $20 \leqslant x \leqslant 100$ (broken line).

The continuous line is the line fitted to the pairs of scores and has the equation $y = 70 + 1{\cdot}5x$, where x is the score on test B and y is the score on test A. The overlap in scores effectively occurs only between 70 and 100 on test A and between 0 and 20 on test B. If we wish to express all possible B scores in terms of A scores we could use the extrapolated broken line giving a total score range on the new common scale from 0 to 220. In this example, we would need to be rather cautious about extrapolating a relationship from such a small part of the score range to cover the whole range, since we have no indication from the data whether the relationship does remain linear outside the range of overlap.

Where the necessary conditions can be satisfied, however, this second solution provides us with a direct method for converting a set of measurement scales to one common scale, which can then be analysed in the same way as other age related measurements. If we adopt the first solution there are some extra restrictions on the kind of analysis which can be undertaken. By defining the mean population scores at each age to be equal, only relative changes between subgroups can be studied. By defining equal variability at each age, these changes will all be measured with respect to the population variability, rather than in terms of any more fundamental scale of measurement.

Finally, we note that an analogous problem occurs when comparing measurements made on similar groups of subjects separated in time. For some types of attitude questions, and for some mental tests, for example, the meaning or interpretation of the questions and tests may change in a population over time. This may affect both longitudinal and secular trend studies. It has been suggested, for example, that inferences concerning changes in the reading ability of English and Welsh children during the 1950's and 1960's are faulty because the use of the same reading tests over a number of years resulted in some of the items being misunderstood through language changes (Burke and Lewis, 1975).

3.4 Age Scale Transformations

Up till now we have used the scale of time or age as a reference scale. This is convenient, since it has precisely the same definition for every individual and in most, although not all populations, can be accurately measured. It thus plays a central role in studies of developmental change.

It is not, however, the only possible scale which extends through time. We might contemplate using height in the same way, for example, both to define the measurement occasions (that is, when predetermined heights are attained), and against which the change in other measurements may be graduated. One limitation of this procedure is that we would have some difficulty in carrying out the measurements at occasions defined in advance. Although it may not seem a very natural scale, it might give a set of very straightforward relationships with other measurements and even suggest important theoretical concepts. This use of a developmental variable as an alternative to age is discussed in detail in the following sections. In the remainder of this section we see how the age scale itself may be transformed.

In Chapter 4 we discuss methods of fitting curves to developmental records. We can view such curves as transformations of the age scale in order to produce a simple relationship between the transformed age and the measurement. For example, if growth in height (y) can be described by the following function of age (t),

$$y = a + bt + c \log t \tag{3.1}$$

then by making the transformation (where we know the values of b and c)

$$x = bt + c \log t \tag{3.2}$$

we get the simple linear relationship

$$y = a + x.$$

With such a transformation, the values of a, b and c vary between individuals so that they must be estimated from a sequence of measurements before we can use the scale. This limits the use of such a scale for purposes such as the construction of standards or the prediction of later development. Nevertheless, such transformations can be useful in the understanding of developmental processes. We shall mention now one age transformation, based on growth curve fitting, which has been widely used.

During adolescence it is well known that, although individuals begin their phase of rapid height growth at different ages and complete it in different periods of time, the shape of their growth curves is similar (Tanner, 1965). Alternatively, if we think in terms of the velocity of height growth, the maximum or peak velocity attained varies between individuals, as does the age at which this maximum velocity occurs. Figure 3.2 shows height velocities for two individuals plotted on the scale of age, and also on a scale where the origin is the age of the individual at which peak height velocity occurs, together with the mean of the two curves.

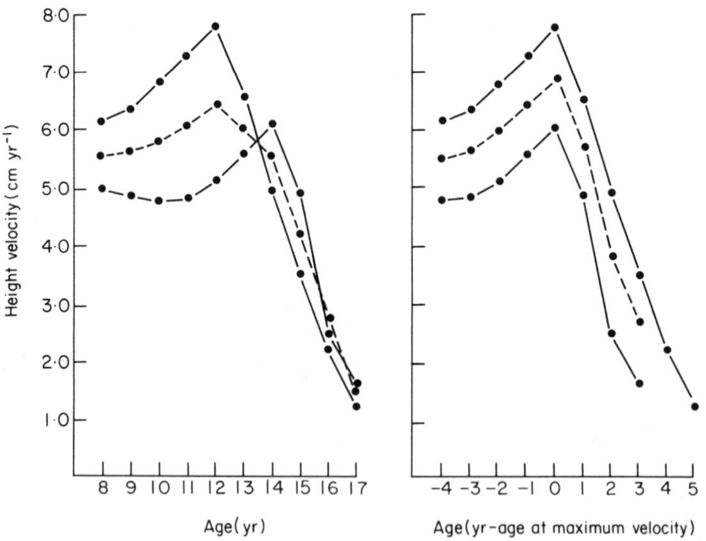

Fig. 3.2 Hypothetical height velocity curves for two individuals (continuous lines) with respect to the age scale, and to the scale of age measured from age at maximum height velocity. The broken line is the mean of the two individual's measurements.

With such a shift in origin, the similar shape of the curves is clear, and we may readily compare the characteristics of individuals on the new scale.

The shape of the curve of the means also reflects more closely the shape of the individual curves.

To produce this transformation a growth curve had to be fitted to height velocity during adolescence in order to estimate the age of peak height velocity (see Chapter 4), and this age then became the origin of the age scale for each individual. Unlike the transformation of equation (3.2) the interval properties of the age scale have been retained, and because these are so familiar, this is an advantage. Naturally, we could use other developmental events in a similar fashion; the occurrence of menarche in girls has often been used in this way.

3.5 Developmental Scales

In order to define a developmental age scale which, in principle at least, will yield a value at any age or time when a measurement may be made, we must look for more or less continuously measurable developmental characteristics. Perhaps the simplest such measurement is one like height, and we shall discuss how height might be used before going on to deal with more commonly used measurements.

We first note that a growing individual's height is monotonically related to his age; that is, as age increases so does height. This is a desirable criterion for a developmental scale and would not be the case with other measurements, such as skinfold thickness or some measures of behaviour. Upon reaching adulthood, however, height ceases to change, whereupon its use as a scale for development disappears. Also, individuals attain different final adult heights, and the scale will therefore have different end points (and starting points) for each individual. We can illustrate this more clearly by making the following scale transformation. For any height, we allocate the age for which this is the mean or median height, according to the appropriate population standards. This is often called the "height-age". For example, a boy with a height of 121·0 cms would be given a height age of 6·9 years according to the 1965 British standards (Tanner *et al.*, 1966). These standards also show that, after about age 17 the median height changes very slowly, and after 18 remains constant at about 175 cm. Thus, not only does a height of 175 cm have a somewhat indeterminate age (we can only say it is 18 or over), those heights greater than 175 cm can be allocated no age at all, even though among, for example, 14-year-old boys over 3% have heights greater than 175 cm.

We can therefore identify two problems with the use of "height age". First, the concept of allocating an age necessarily becomes vague as we near the end of the developmental period and the measured characteristic ceases to change. This is an inherent problem of age-related developmental scales which we shall discuss further below. Secondly, a measurement such as height reflects not only the level of development, but also the final size which will be attained by the individual. If we wish to eliminate this latter problem we could define relative height. This immediately leads us back, however, to defining a measurement which cannot be used until development has finished, and must therefore be ruled out of the present discussion.

3.6 Maturity Scales

We now turn to a different approach to the problem by developing the idea of a maturity scale. The notion of maturity supposes two well-defined states, namely complete immaturity and complete maturity. Thus, a scale of maturity has a defined beginning and end point, and the scale is traversed, albeit at different rates, by every normal individual during development. Although the definition and measurement of maturity can be applied to many different aspects of development, for purposes of illustration we shall use the much studied example of skeletal maturity scales, the main elements of which can be summarised as follows.

The carpal centres and the epiphyses of the hand bones become visible in radiographs taken during late fetal life, and each bone passes through a recognizable series of stages until it assumes its final adult shape, usually by 20 years of age. While the growth of the bones themselves is continuous, up to about nine recognizable stages can be defined for each one, so that in an individual the stages of any one bone occur in a given sequence. These stages are defined according to their shapes, using criteria which are independent of their absolute size, since this may itself be related to the size of the individual. This provides the basis for a measurement of maturity, which is then made by combining the information from the 28 separate bones into a single measure of the presumed underlying continuum known as skeletal maturity.

The following criteria are necessary for a set of characteristics to reflect a particular area of maturity. We shall refer to these characteristics as "indicators" of maturity.

(i) Each indicator is measured on an ordinal scale with a fixed starting point and a fixed end point. If a continuous measurement is available, this can be suitably grouped to provide ordered categories.
(ii) Intermediate values or stages of each indicator have a well defined ordering according to age, and the ordering is the same for all individuals.
(iii) Each measurement is made in a way which is independent of the individual's size.

(iv) Each indicator is fully defined at any occasion, without reference to measurements at other occasions.
(v) It is assumed that there is a continuous underlying dimension, known as the maturity, of which each indicator is a manifestation, and that these indicators are to be combined in some fashion to give an estimate of the maturity.
(vi) Every normal individual passes through every stage of every indicator. By "normal" we mean those individuals in any given application who are deemed to constitute a suitable standardizing group.

Strictly speaking, it is the set of measurements we choose which *defines* the maturity scale, just as it is the set of items in a IQ test which defines "intelligence".

3.6.1 Age Prediction Methods

Historically, the measurement of maturity was first approached by looking for a combination of maturity indicators which was a good predictor of age. As we shall see, this approach presents practical as well as conceptual difficulties and in the next section we shall present an alternative formulation which avoids the difficulties.

One of the best known methods of measuring skeletal maturity is the "Atlas" technique, developed by Greulich and Pyle (1959). This consists of pictures of ordered individual wrist radiographs, each one selected as typical of an age. An individual is then assigned the age corresponding to the picture which most closely resembles the set of bone stages exhibited by his own radiograph. Since only a few of all the possible pictures can be displayed in the Atlas, subjective judgements of similarity have to be made, possibly leading to consistent differences between measurers. (In fact, Greulich and Pyle use 31 pictures for boys and 29 for girls, out of the very much larger total number which are seen in practice.)

We can see that this is essentially a method of predicting age from maturity indicators, by noting that the method assigns an age to each picture, which is in fact an estimate of the average age of all individuals in the population who exhibit that picture. If the predicted age for an individual is, say, much less than his chronological age, we may assert that his skeletal development is typical of individuals of a much younger age or, more simply, is delayed.

Other age prediction methods have been suggested, which avoid the subjectivity of the Atlas technique. The simplest such methods (see, e.g. Wolanski, 1966 and Moorees *et al.*, 1963) give each stage of each bone a score which is proportional to the mean age of the individuals who exhibit the stage, and then average these scores. (Needless to say, there is a subjective element which remains in this, as in all other methods, namely that of identifying the appropriate stage from a radiograph.) Although it does not appear to have been done, it would be possible to improve on this method by using a weighted average of the individual stage scores to predict age. For example, we could carry out a multiple regression with age as the dependent variable and the bones with their stage scores as independent or predictor variables.

A slightly different approach to age prediction is to assign ages to the *transitions* from one stage to another, rather than to the presence of an individual in a stage. This has usually been done by using probit or logit techniques for estimating average ages of transition, in a similar way to that described in Chapter 5 for estimating the average age of menarche. Knowing the mean or median ages of transition, and assuming that the variability about the age of transition is the same for all transitions within a bone, we may then designate the mean of two adjacent transition ages as the age associated with being in the corresponding stage. As before, individual's skeletal age may be estimated as the average of these ages. It appears that this procedure is merely a rather complicated version of the one described in the preceding paragraph.

There are three general problems associated with all the age prediction methods. Firstly, arbitrary decisions must be made about the ages associated with complete immaturity (with all the bones exhibiting the initial stage) and complete maturity (with all the bones exhibiting the final stage). Since, by the age of about 20 years, all normal individuals are fully mature, any age after this could be said to be "typical" of a fully mature radiograph. The mean age associated with such a radiograph will therefore depend on the upper age which we use to define our population; the older this becomes the higher becomes the average age associated with a fully mature radiograph. It is because we have no transition out of the final stages that we are forced to impose an arbitrary upper age, and likewise for the initial stages.

Secondly, by using age to define the maturity scale, we may be obscuring an underlying biological dimension of maturity which has a quite different relationship to age.

Finally, the ages associated with the stages, and thus the ages predicted from given combinations of stages, are known to vary over time and from population to population. This means that we can have no common measurement scale, a fact which imposes a limitation on comparative studies.

3.6.2 Scaling Methods

Acheson (1954, 1957) proposed a method of scoring bone stages which is independent of age, and relies only upon assumptions about the contribution which each stage should make towards the overall maturity score. He numbered the successive stages of each bone, 0, 1, 2, ..., and estimated skeletal maturity as the mean of these scores for an individual. Such a procedure for scaling responses or categories is a familiar one in the social and behavioural sciences, and an account of various scaling methods can be found in Torgerson (1958). The one which we discuss below is a modified version of that first described by Guttman (1941).

The basic assumption behind Acheson's system is that, passing from any one stage to the next in any bone contributes the same amount to the overall maturity. In general, however, this assumption constitutes an oversimplification and we need to search for a more realistic scoring system. With suitable theoretical knowledge we might be able to define an appropriate scoring system, but since this seems rather unlikely at present in most applications, we turn to another approach.

As we have already stated, we can define two fixed points on any maturity scale, namely the starting point of complete immaturity and the end point of complete maturity. The actual numerical values we give to these points of the scale are immaterial, and one convenient choice is 0 and 100 respectively. We shall illustrate the problem of assigning scores to the individual stages, which can then be averaged to provide maturity scores, by discussing a very simple system with just two bones, each with two stages.

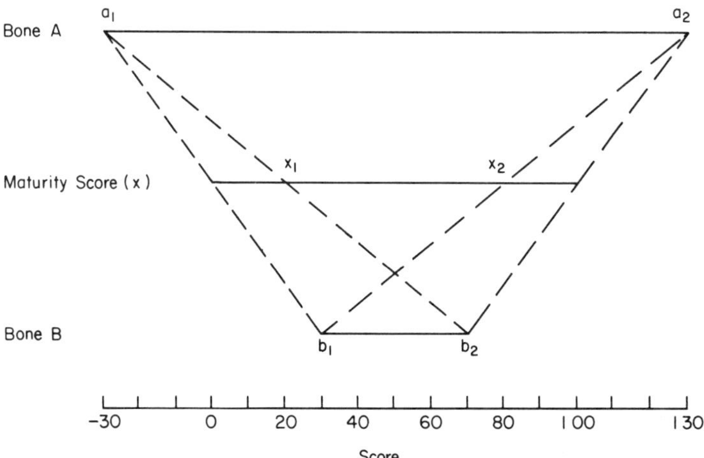

Fig. 3.3 Hypothetical illustration of maturity scoring procedure. The three score scales for Bone A, Maturity Score, and Bone B are calibrated by the lower horizontal scale.

In Fig. 3.3 the two continuous horizontal lines represent scales for two bones A, B and the line halfway between is the simple mean, and represents the maturity scale (x), extending from 0 to 100. The points a_1, a_2, b_1, b_2 represent one possible set of scores for the two stages on each bone. The line joining $a_1(-30)$ and $b_1(+30)$ gives the average value of these two points where it crosses the broken line, namely zero. Likewise the line joining $a_2(120)$ and $b_2(80)$ crosses the broken line at the point with score

100. The point $x_1(25)$ gives the maturity score for an individual exhibiting the first stage of bone A and the second stage of bone B, and vice-versa for the point $x_2(75)$. These four maturity scores, in fact, exhaust all the possibilities in this simple example.

We recall that one of our assumptions is that each bone reflects the same underlying attribute of maturity. Hence, we would expect, in a typical individual, that since each of the scores for the observed stages is estimating the same quantity, they should be close to each other in value. Using a large standardizing sample of subjects in order to estimate the scores for the stages, we are therefore led to adopt a criterion for determining the best set of scores which is based on minimizing the disagreement between the scores assigned to the observed stages for each individual subject.

For simplicity, let us measure disagreement by the simple difference in score (ignoring its sign or direction) between the two observed stages in an individual. Thus, an individual with a score of 30 units on bone A and 50 on bone B would generate a disagreement of 20 units. Now consider the two possible alternative scoring systems set out in Table 3.3. Note that the end constraints are satisfied by both systems.

Table 3.3 Two hypothetical scoring systems

Stage	System	
	1	2
a_1	–1	–20
a_2	115	160
b_1	1	20
b_2	85	40

In Table 3.4 we give the individual disagreement generated by these scoring systems for each possible combination of stages, together with the distribution of a hypothetical sample of 100 individuals. The overall disagreement is the sum of the individual disagreements.

We see that, with the same sample of individuals, system 1 generates less overall disagreement than system 2, and hence is to be preferred. By suitably varying the positions of the points *a*, *b*, *c*, *d*, subject to our end constraints, we can determine the set of scores which gives the smallest possible overall disagreement, and which, according to our criterion, is the best scoring system. In practice, because it simplifies the algebra, a slightly different definition of disagreement is used, based on the sum of the squared differences rather than the absolute differences between scores. The extension of the principle to more stages and bones is straightforward. We can also add a refinement, by differentially weighting the bones so that some bones play a greater role in determining the overall maturity score. A full mathematical description is given by Healy and Goldstein (1976), and the application to the bones of the wrist is described by Tanner *et al.* (1975). These authors modify the scoring system very slightly by requiring the *sum* of the stage scores to lie between 0 and 100 and add a suitable constant to the scores for each bone in order to avoid negative stage scores.

Table 3.4 Stage combinations, maturity scores and disagreements for scoring systems of Table 3.3

Stage combination	Maturity Score	Disagreement (d)	Number of individuals exhibiting each combination (n)	Total disagreement (nd)
System 1				
a_1b_1	0	2	50	100
a_1b_2	42	86	15	1290
a_2b_1	58	114	10	1140
a_2b_2	100	30	25	750
		Overall disagreement = 3280		
System 2				
a_1b_1	0	40	50	2000
a_1b_2	10	80	15	900
a_2b_1	90	140	10	1400
a_2b_2	100	120	25	3000
		Overall disagreement = 7300		

The method we have described defines maturity without reference to the age scale. It does not make arbitrary assumptions concerning differences between successive stages, and the indeterminacy at the ends of the scale which characterized the age prediction methods, no longer applies, since the ends have been assigned fixed values. It is potentially applicable to different populations at different times, although whether the scoring system remains invariant is a question to be settled by empirical studies.

A measurement of maturity, by indicating how far an individual has travelled along the path to full maturity, may contribute useful information for the prediction of the mature values of other measurements. Tanner *et al.* (1975)

show how this works for the prediction of adult height using measurements of skeletal maturity and height at particular ages.

One further point is worth emphasizing. The scores will depend on which bones or attributes it is decided to incorporate into the system. By selecting a particular group of bones we are defining thereby the aspect of maturity to be measured, and we must make such a decision on the basis of whatever evidence or theory we have available. Once having decided, say, to use 20 bones of the hand and wrist, we can study the question of whether excluding particular bones or amalgamating stages actually produces an important difference in the set of overall maturity scores. Since, in general, there are no external criteria by which to judge the size of such a difference, we may only suggest the following general principles.

The first of these states that the number of bones and stages should be sufficiently large to guarantee a smooth relationship between average maturity and age, reflecting the undoubtedly continuous nature of maturity itself. The second principle states that the variation of maturity scores between the individuals in a population at given ages, should be large enough to reflect the undoubtedly considerable variations in maturity between individuals. Finally, we can contrast two systems satisfying both these principles, by comparing the pairs of maturity scores given by the two systems, for the individuals in a standardizing sample. An example of the practical application of these principles is given in Demirjian and Goldstein (1976).

3.6.3 Sampling Considerations

One problem still remains before we can set up our maturity scale, namely the selection of a suitable sample from which to obtain the score estimates. First, we must define a reference population from which we require a representative, preferably random, sample and to which our scoring system will be applied. We assume that the physical characteristics of geography, demography etc. have been settled and we are left with defining the appropriate age range.

It seems fairly obvious that, over the age range in which we are interested, all ages should be equally represented. This implies, however, that we must determine some cut-off points to avoid including large proportions of wholly immature or wholly mature individuals. It is clear from the example of the previous section that the more of such individuals who are included, the more nearly equal this will force the set of initial stages and also the set of final stages. Thus, we are faced with the somewhat undesirable feature that the scoring system will depend on our choice of age range. It seems that we must accept some arbitrariness of this kind, although in practice this may not be a serious problem.

The sample size necessary to provide reasonably accurate estimates of stage and individual scores will vary according to the number of indicators and stages. For a skeletal maturity system with about 20 indicators and nearly 200 stages, between 3000 and 4000 sets of observations seem to be adequate. If we wish to construct percentile standards from the estimated individual scores, the sample sizes necessary to achieve a given precision are those given in Chapter 6. It will often be the case that the same sample is used, both to estimate stage scores and to provide standards over a wide age range. In this case the total sample size to obtain acceptable accuracy for percentile estimates may be larger than that required to obtain acceptable accuracy for the scores. Also, however, as shown in Chapter 6, for percentile estimation the sample should not be spread uniformly over the age range, whereas it should be in order to estimate the scores. In such a case we can proceed by sampling for the purposes of percentile estimation, and in order to estimate the scores, carry out the basic calculations within narrowly defined age groups and then combine the age groups, suitably weighted, in order to simulate a uniform age sample (see Tanner *et al.*, 1975).

3.6.4 Other Applications

Although we have dwelt on the development of a skeletal maturity system, this is only because most work has been done with that system. In principle, there is no reason why a similar notion of maturity should not be applied to other areas, such as behavioural or cognitive development. For example, our concept of maturity indicators seems to be implicit in much of the work of Piaget (Tanner and Inhelder, 1969), who has defined stages of mental development formally analogous to those of bone development. There also seems to be no reason why certain aspects of reading or mathematics development should not be defined in terms of relative maturity. If we set developmental objectives, in terms of aspects of language for example, it may be possible to define recognizable stages through which all normal individuals will pass. An area where the maturity approach would appear to have particular application is in so-called developmental screening or testing. Typically, these procedures consist of large numbers of items describing whether or not a child exhibits certain behaviours. Distinct areas of development are often defined, such as "fine motor ability", and the concept of relative maturity seems to be assumed implicitly.

References

Acheson, R. M. (1954). A method of assessing skeletal maturity from radiographs. A report from the Oxford Child Health Survey, *Journal of Anatomy*, **88**, 498–508.

Acheson, R. M. (1957). The Oxford method of assessing clinical maturity, *Clinical Orthopaedics*, **10**, 19–39.

Bock, R. D. (1976). "Multivariate Statistical Methods in Behavioural Research", McGraw Hill, New York.

Burke, E. and Lewis, D. G. (1975). Standards of reading: a critical review of some recent studies, *Educational Research*, **17**, 163–174.

Cameron, N. (1979). The methods of auxological anthropometry, *In* "Human Growth: A Comprehensive Treatise" Vol. 1 (Faulkner, F. and Tanner, J. M. eds) Plenum Press, New York.

Demirjian, A. and Goldstein, H. (1976). New systems for dental maturity based on seven and four teeth, *Annals of Human Biology*, **3**, 411–421.

Duncan, O. D., Schuman, H. and Duncan, B. (1973). "Social Change in a Metropolitan Community", Russel Sage Foundation, New York.

Goldstein, H. and Blinkhorn, S. (1977). Monitering educational standards—an inappropriate model, *Bulletin of the British Psychological Society*, **30**, 309–311.

Greulich, W. W. and Pyle, S. I. (1959). "Radiographic Atlas of Skeletal Development of Hand and Wrist", 2nd edition, Stanford University Press, California.

Guttman, L. (1941). The qualification of a class of attributes; a theory and method scale construction, *In* "The Prediction of Personal Adjustment" (Horst, P. ed) New York Social Sciences Research Council.

Healy, M. J. R. and Goldstein, H. (1976). An approach to the scaling of categorised attributes, *Biometrika*, **63**, 219–229.

Hoinville, G. and Jowell, R. (1978). "Survey Research Practice" Heinemann Educational Books, London.

Moorees, C. F. A., Fanning, E. A. and Hunt, E. E. Jr (1963). Age variation of formation stages for ten permanent teeth, *Journal of Dental Research*, **42**, 1490–1502.

Moser, C. and Kalton, G. (1971). "Survey Methods in Social Investigation", Heinemann, London.

Tanner, J. M. (1962). "Growth at Adolescence", Blackwell, Oxford.

Tanner, J. M., Whitehouse, R. H. and Takaishi, M. (1966). Standards from birth to maturity for height, weight, height velocity and weight velocity: British children 1965. II, *Archives of Diseases in Childhood*, **41**, 613–635.

Tanner, J. M. and Inhelder, B. (1969). "Discussions on Child Development", Tavistock Publications, London.

Tanner, J. M., Whitehouse, R. H., Marshall, W. A., Healy, M. J. R. and Goldstein, H. (1975). "Assessment of Skeletal Maturity and Prediction of Adult Height: TW2 Method", Academic Press, London and New York.

Thorndike, R. L. (ed) (1971). "Educational Measurement", 2nd edition, American Council on Education, Washington, DC.

Torgerson, W. S. (1958). "Theory and Methods of Scaling", Wiley, New York.

Wolanski, N. (1966). A new method for the evaluation of tooth formation, *Acta Genetica et Statistica Medica, Basle,* **16**, 186–197.

World Health Organisation (1975). "World Health Statistics Annual".

4

The Analysis of Models for Time-related Measurements

This chapter deals with models where the same measurement is made on one or more individuals on a number of occasions. For example, we can readily make a measurement of weight at successive ages and effectively the same measuring instrument is involved each time. More specifically, these models are concerned with the manner in which a measurement is related to age (or another time scale), for example by studying its average rate of change with age. The models studied in this chapter relate to continuous measurements. Analogous models for discrete measurements are dealt with in section 5.6.4 of Chapter 5. Chapter 5, however, is mainly concerned with the other principal type of model applicable to longitudinal data which deals with the relationships between the value of measurements on one specified occasion and the value of the same or different measurements on other specified occasions.

We begin by tracing the historical development of models fitted to human body growth measurements.

4.1 Fitting Growth Curves to Individual Records

One of the earliest descriptions of a pattern of growth is the record kept in the 18th century by Count de Montbeillard of the growth of his son (Scammon, 1927).

We note two things from Fig. 4.1 which are also characteristic of a large number of other types of growth. Firstly, apart from small fluctuations, the pattern of growth follows a fairly smooth curve, and secondly the actual shape of the curve is complex. The former observation has encouraged researchers to attempt to find relatively simple mathematical descriptions of growth, whilst the latter observation has made this task extremely difficult. If we study the "velocity" or "rate of change" plot, the continuous line in Fig. 4.2, we can see why this is so. This curve highlights details of the growth pattern—a high velocity immediately after birth, falling to the age of four years, then fairly constant and falling again from the age of eight years to 12 years—rising to age 15 years followed by a decline to zero at age 18 years when adult height is reached. It would be possible to draw a smooth mathematical curve which followed this pattern more or less closely. For example, a simple cubic curve would fit the observed data rather poorly, whereas a sixth degree polynomial fits better but still does not follow the observed pattern very closely.

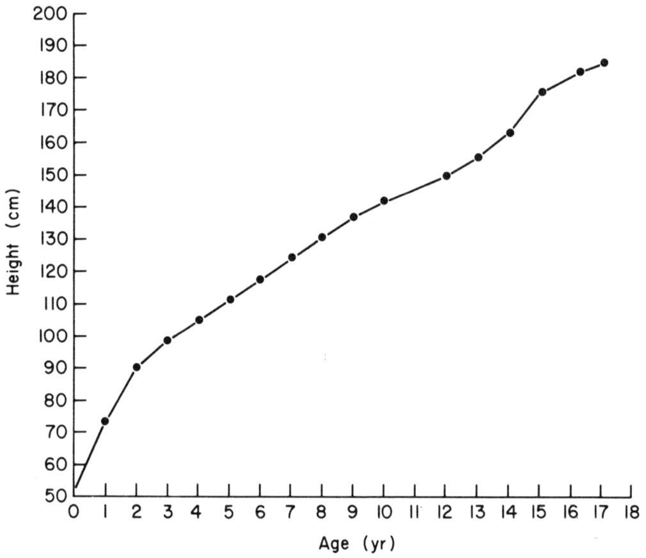

Fig. 4.1 Height measurements from birth to 17 years on de Montbeillard's son (1759–1776). Only points at or near birthdays are plotted in order to avoid short term seasonal fluctuations (see Chapter 2).

By fitting higher order polynomials containing more and more coefficients, we can follow the actual observed plot more and more closely. The need is to find a balance between a close fit, and a simple summary which captures the essential characteristics of growth. We shall return to a detailed discussion of this after discussing various attempts to tackle the problem.

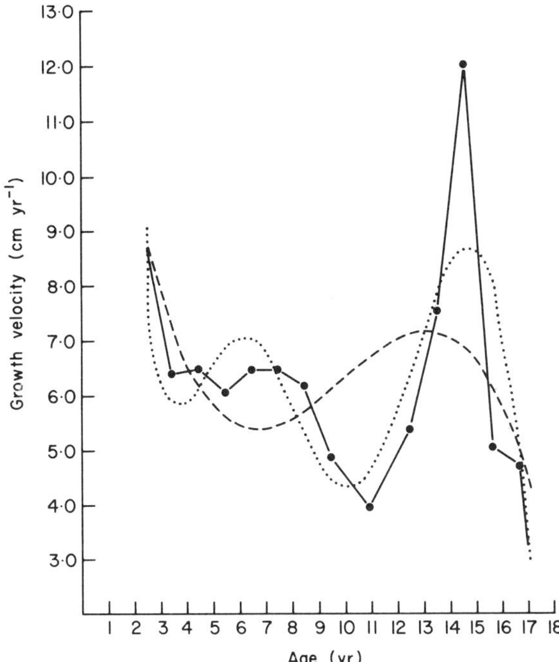

Fig. 4.2 Height velocity measurements from birth to 17 years on de Montbeillard's son. Velocities are calculated from measurements taken one year apart. The continuous line represents observed velocities calculated by simple subtraction. Dashed line represents fitted cubic curve. Dotted line represents fitted 6 degree curve.

4.1.1 Single Repeated Measurements: An Historical Survey

Towards the end of the 19th century, large numbers of measurements, mainly of height and weight, began to be accumulated on children of different ages, and curves of growth and percentile standards had been constructed from these cross-sectional measurements. It was suggested by Boas (1892), however, that through the medium of poverty, childhood mortality was associated with growth attained and was high enough to lead to serious biases if one wished to use the cross-sectional data to obtain an estimate of the average growth curves of children. He pointed out that the best way to study actual growth was to obtain repeated measurements on the same individual, and he suggested that longitudinal studies should be set up. The particular considerations which led Boas to this conclusion might not apply to the same extent today although factors such as differential geographical mobility may play just as important a part, a point already discussed in Chapter 2.

An important development in the mathematical description of growth came with Robertson's (1903, 1915) development of his "autocatalytic" equation which he fitted to mean weights of babies up to 12 months old. Robertson argued that the percentage rate of change of size decreased proportionately to the size itself. Percentage rate of change is measured by the relative growth rate, i.e. the growth velocity divided by the size. Thus when the size is very small, the relative growth rate is high, but the actual growth is low because of the small size. As size increases the growth rate increases and when size nears the final size, the relative and actual growth rates slow down.

If we write for the growth rate

$$\frac{\mathrm{d}y}{\mathrm{d}t} = \frac{b}{k}y(k - y) \tag{4.1}$$

where k equals the final size, we can see what this implies for growth.

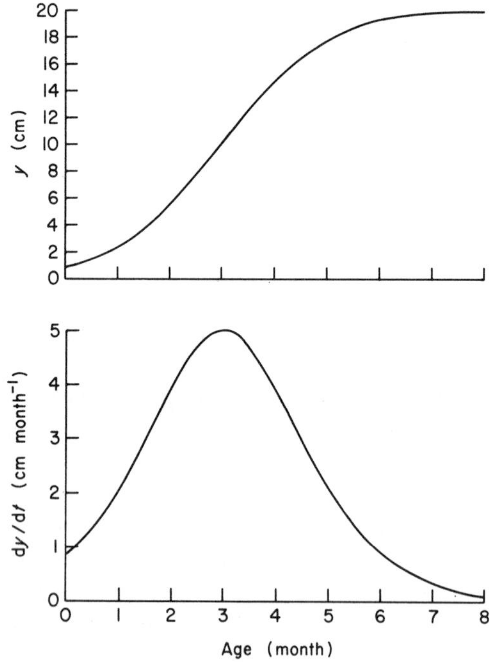

Fig. 4.3 Hypothetical growth curve of the length of an animal in cm (y), with equation $y = 20/(1 + e^{(3-t)})$ and $\mathrm{d}y/\mathrm{d}t = \frac{y}{20}(20 - y)$.

When y is small relative to k, then the growth rate is small. When y equals $k/2$, the growth rate is fastest and as y approaches k, the growth rate again tends to zero. The graph of growth rate against age is symmetrical about the value $y = k/2$, as can be seen in the upper part of Fig. 4.3 which represents a hypothetical curve of the length of an animal over the age range 0–6 months.

The lower part of Fig. 4.3 shows the "distance" curve of y against age (t), given by the following equation for y, which satisfies the growth rate eqn. (4.1),

$$y = \frac{k}{1 + e^{a - bt}}. \tag{4.2}$$

This has a shape typical of the growth of many organisms and parts of organisms. It is symmetrical about the point of maximum growth velocity, here where $y_t = 10$ cm and $t = 3$ months. It can be seen from Fig. 4.3 that as t becomes very large or very small, so the value of y tends slowly to approach the values k or 0, these values being known as the upper and lower asymptotes. In the case of most growth measurements the upper asymptote may be thought of as final adult size, which is approached slowly at the end of the growing period. An interpretation of the lower asymptote is often more problematical since growth rarely starts slowly from zero or near zero size. However, by the addition of a positive constant to the above equation we may obtain a reasonable approximation to a part of a growth cycle. Writing

$$y = c + \frac{k}{1 + e^{a - bt}}$$

we have a lower asymptote for the value $y = c$, and an upper asymptote for $y = c + k$.

This curve, known as the logistic, has played a central role in the history of the subject, and we shall return to it later when discussing more recent attempts at fitting growth curves to individual growth records.

In the next two or three decades, various modifications to the basic logistic or autocatalytic equation were suggested. Pearl and Reed (1923), for example, replaced the expression $(a - bt)$ by a general polynomial in powers of t, thus introducing more unknown parameters and making the curve more flexible. A curve very similar to the logistic is the Gompertz curve, first suggested for use as a growth curve by Wright (1926) and developed by Winsor (1932). The equation for this can be written

$$y = k \exp(-\exp(a - bt)). \tag{4.3}$$

It has the same general shape as the logistic, and became a popular curve to fit to human and animal growth data. In part it seems that this popularity can be attributed to a simple interpretation of the equation. The relative growth rate is given by

$$\frac{1}{y}\frac{\mathrm{d}y}{\mathrm{d}t} = b\mathrm{e}^{a-bt}.$$

Thus the relative growth rate decreases exponentially with time, that is in equal small intervals of time there are equal proportionate decreases in the relative growth rate.

Various other workers experimented with other types of curves. Jenss and Bayley (1937) found that the following curve was a good fit for growth in height for children up to about six years.

$$y = a + bt - \mathrm{e}^{c+dt}.$$

The logistic and Gompertz, however, found the widest use, both because of their relative simplicity and because of their applicability. For human growth they are used extensively to fit growth in height during adolescence.

One of the problems encountered in fitting such curves to only part of the age range, is to know precisely which ages to include. For example, a fitted curve can have different parameter values if extended to include measurements made at earlier ages. Furthermore, at least in human growth, children enter the adolescent phase of growth at different ages, and a decision as to when this occurs has to be made for each child. Marubini *et al.* (1972) use a preliminary examination of the velocity curve for an individual to decide which ages to include, choosing that age at which the velocity prior to adolescence takes a minimum value. This procedure is rather like having an extra parameter value to estimate, and we will return to the question of the number of parameters below. In order to avoid this problem and to obtain a curve covering the whole growth period, several authors have proposed to describe growth as the sum of several logistic or Gompertz curves. Burt (1937), for example, suggested that the growth in height of girls from birth to 18 years could be represented as the sum of three logistic curves. Zucher *et al.* (1941) applied growth curves of this type to studies of rat growth.

An alternative approach would be to fit separate curves for each age period, with a suitable method for establishing their point of joining and for providing that this should be a smooth transition. Following this approach Deming (1957) proposed the curve studied by Jenss and Bayley (1937) followed by the Gompertz for growth in height of children. Unfortunately, however, the problem of suitably joining the curves was not satisfactorily solved. It was with the advent of electronic computers that the extensive computation necessary for empirical comparison of different types of curve could be undertaken. Before dealing with that development, however, we shall mention a somewhat different approach.

As well as being interested in fitting curves such as the logistic in order to graduate, or closely follow, observed measurements with smooth curves, some workers developed models of growth based on a presumed understanding of underlying physiological processes. They attempted to relate these processes to growth in terms of mathematical formulae, arriving eventually at growth curves whose form reflected the supposed underlying processes, although necessarily still remaining oversimplifications of the actual underlying processes. Weiss and Kavanu (1957), for example, derived a 14 parameter equation to explain the growth of chickens, each parameter having a meaningful physiological interpretation. Their physiological model is complex, involving feedback mechanisms, but they do fit it to empirical data and give interpretations of the parameter estimates they obtain. There appears to have been little attempt to follow up this approach, partly because of computational complexities but also because many growth records have only a relatively small number of measurements, so that multi-parameter curves may graduate growth well but provide little accuracy for the estimated parameters.

Once it had become possible to carry out the routine fitting of growth curves to data, this stimulated the comparison of different types of curves and also the development of some more general types. The Gompertz and logistic curves were shown to be special cases of a more general logistic type curve (Nelder, 1961, 1962), and several workers compared this type of curve with a basic polynomial curve. Kidwell and Howard (1970), compared the fitting of Gompertz and second degree polynomial curves to weight growth of mice in the first 10 weeks of life. Both groups of curves contain three parameters and the authors conclude that in terms of goodness of fit the polynomial curve is somewhat more satisfactory and for the purposes of genetic studies comparing the parameters of individual curves, neither appears to have a clear advantage. They point out that the parameters of the Gompertz have readily understood interpretations in terms of maximum growth rate and also that a polynomial will provide convenient summaries of growth such as the average growth rate. Since the aim of much growth curve fitting is to produce a summary of growth in terms of a few parameters, it is reasonable to judge the results by the utility of these summaries. With suitable computer programs many sets of data can be fitted rapidly and the resulting parameter values used to compare, say, groups of individuals. Bock *et al.* (1973) fit a sum of two logistic curves to the successive height measurements of each child in a sample of over 100 children. Their equation is

$$y = \frac{a_1}{1 + \exp(b_1 - c_1 t)} + \frac{F - a_1}{1 + \exp(b_2 - c_2 t)}$$

where F is the final size attained, and is assumed to be known.

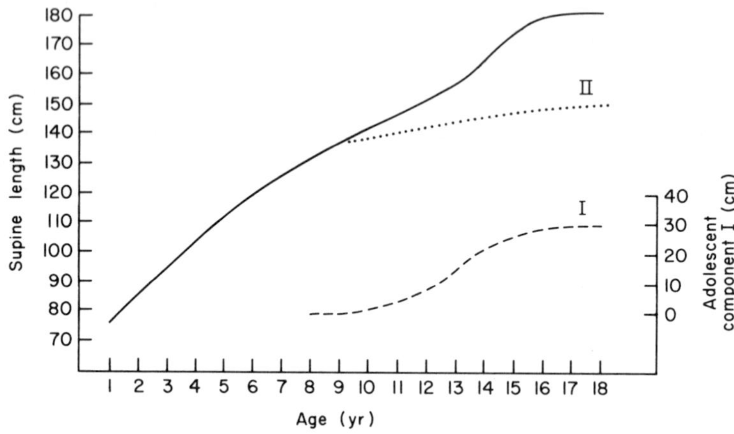

Fig. 4.4 A two component logistic fitted to supine length for a single boy. (Data from Bock *et al.* (1973).) The pre-adolescent (II) and adolescent (I) component are also shown separately. The equation is

$$y = \frac{151{\cdot}5}{1 + e^{(0{\cdot}25 - 0{\cdot}27t)}} + \frac{28{\cdot}9}{1 + e^{(12{\cdot}84 - 1{\cdot}01t)}}$$

It can be seen clearly in Fig. 4.4 how the second component begins at about the age of nine years, with a relatively small change in the first component after that age. These authors do not attempt to fit the equation below one year, but do claim that from one year to adulthood a satisfactory fit is obtained. It is of interest to note, however, that there tends to be a systematic fluctuation of observed points about the fitted curve, e.g. falling below the line in the range four to six years and above in early adolescence. This indicates that there is a systematic element of growth not being explained. Such trends have not been noted by all investigators. Tanner *et al.* (1976), for example, find no such systematic element when they fit a single four-parameter logistic to adolescent growth. Bock *et al.* (1973) go on to make comparisons between the sexes in terms of the parameters of the fitted curves and also suggest that because of high correlations between them, only four of the six parameters are necessary for an adequate description of growth in height. Preece and Baines (1978) suggest another set of mathematical curves which use five or six parameters, one of which is the final size attained, which is not therefore assumed to be known in advance. Although no simple interpretation can be given to all the individual parameters in terms of separate components of growth, the curves do give better fits than those of Bock *et al.* (1973) when fitted to the heights of about 60 children with measurements from about four to 18 years. Nevertheless, systematic fluctuations are still present and not all events are estimated accurately, in particular the value of the maximum height velocity. Lozy (1978) points out that the logistic curves give generally poor estimates of the velocity curve. In another application, Bock and Thissen (1976) fit the sum of three logistic curves to height measurements from one to 18 years and these appear to give more satisfactory estimates of the velocity curve, although this is at the cost of having to estimate a total of eight parameters. All these authors describe computational methods of fitting the curves.

Another approach, which seems not to have been widely used, is to fit separate curves to different sets of occasions with smooth joins where the curves meet. For example, the Jenss and Bayley curve might be joined to a logistic curve at about age 10. The precise age would differ between children, the optimum join point being determined as that which gave the best fit, say by minimizing the residual sum of squared deviations from the fitted lines. The join point would also then be of interest as an individual characteristic. Hudson (1966) discusses such piecewise procedures for carrying out the curve fitting.

Another approach is to use so-called spline functions (see e.g. Poirier, 1973) which are sets of polynomials (usually at most cubic curves) joined together smoothly.

We shall now discuss some of the general problems of selecting suitable growth curve models by considering the problem of fitting curves to human adolescent growth in height. We shall suppose that the age range has been selected and includes a good estimate of adult size. Adolescent growth has certain properties and any curve has to incorporate these. For example, consider the ratio of height at the time of maximum velocity to final adult size, both measured from the height at the start of adolescent growth. For the Gompertz this is fixed by the nature of the curve at 0·37 and for the logistic at 0·50. For adolescent growth, however, the average value calculated empirically, seems to be about 0·40, with most individuals falling in the range 0·35–0·45 (Frisch and Revelle, 1969; Deming, 1957). Hence we cannot expect these curves to estimate accurately both the height at the time of maximum velocity and the final adult height for all individuals, although we would expect the Gompertz to do better than the logistic. It is characteristic of growth curves derived from models similar to equation (4.1) that several numerical relationships between growth events of interest are fixed by the form of curve.

As pointed out earlier, one solution to this problem is to seek more general curves which have more parameters and can therefore be made to fit observed growth more closely. Of course, it would be possible to fit observed growth exactly, simply by fitting a polynomial curve of degree one less than the number of occasions. The aim, however, is not just to graduate growth as closely as possible, but to do so with as small a number of parameters as possible which summarise the salient features. Thus, for this purpose the short-term irregularities of growth are of no interest and the function of the growth curve is to smooth these out. The generalization of the simple logistic curve referred to earlier and given by Nelder (1961, 1962) is

$$y = \frac{k}{(1 + ce^{a - bt})^{1/c}}.$$

For this curve the value of y at the time of maximum growth velocity is $\frac{k}{(1 + c)^{1/c}}$, so that individual variation in the ratio of this value to adult height will determine the additional parameter c and this gives us some flexibility in accommodating empirical relationships. There appears to have been little systematic exploration of the use of generalizations of the more commonly used curves such as the above. One of the practical problems is that as the number of parameters increases the accuracy with which each is measured decreases and as Bock *et al.* (1973) found, the correlations between the parameters become high, making interpretation difficult.

An alternative to seeking more flexible general systems of the above type, is to adopt a more direct approach to estimating the events of interest. If, for example, we wish to estimate the age at maximum growth velocity, then we might select by eye the set of four or five occasions which contain this age, and fit a simple second or third degree polynomial to pass through these points. We could, of course, fit a higher degree polynomial to pass through all the points but since the measurements themselves have some error attached to them, it is more appropriate to allow this to be reflected in the variation about a lower order curve which will be more stable. The age when the fitted curve has maximum slope is then easily determined by differentiation. Figure 4.5 shows a second degree and a third degree curve fitted respectively to five points for De Montbeillard's son around the age of 14 years.

Clearly the quadratic curve, which departs only slightly from a straight line, loses the essential shape of the growth curve and is of little use for estimating the maximum velocity. The cubic curve gives a much better fit and gives an estimated maximum velocity of 10·7 cm yr^{-1} at age 14·17 years. This compares with the maximum velocity estimated directly from the consecutive yearly increments in height of 12·1 cm yr^{-1} (Fig. 4.2). We note that the estimated maximum velocity of 10·7 cm yr^{-1} is the instantaneous velocity at a single age. In fact, we are often more interested in the maximum velocity for consecutive measurements a year apart, as when compiling growth velocity standards, and it would then be appropriate to start from the consecutive yearly differences between such measurements, such as are plotted in Fig. 4.2. If we do this and fit a quadratic to the four velocities derived from the above five measurements we arrive at a maximum velocity of 10·5 cm yr^{-1}.

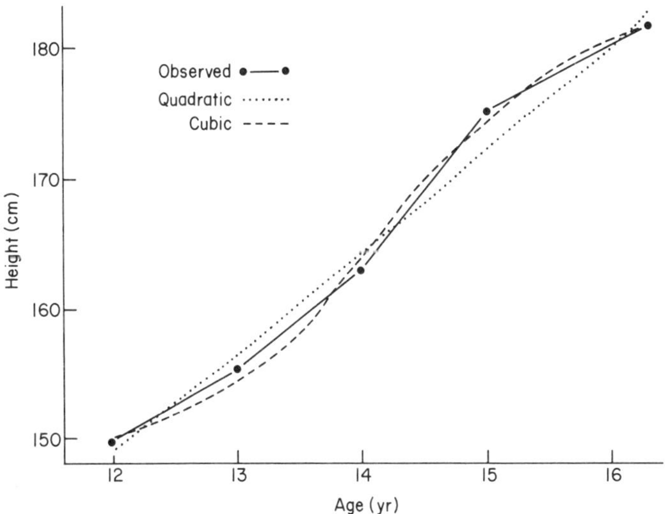

Fig. 4.5 Quadratic and cubic curves fitted to de Montbeillard's son's height measurements from 12 to 16 years.

Quadratic equation is $h = 68{\cdot}6 + 5{\cdot}86t + 0{\cdot}069t^2$.
Cubic equation is $h = 2097{\cdot}2 - 430{\cdot}4t + 31{\cdot}13t^2 - 0{\cdot}732t^3$.

Finally, therefore, it would appear that while reasonably close graduation of individual growth often can be achieved with relatively simple curves, these do not summarize all aspects of interest nor do they necessarily fit all individuals. Furthermore, the form of equation used itself determines certain fixed relationships which may not hold for many individuals. For some purposes, the parameters of such curves may provide useful summaries which, for example, highlight group differences. For other purposes it may be more appropriate to use special techniques for particular events of interest, fitting simple polynomial equations over a restricted age range. In experimental studies especially, the relationship between the growth measurement and time may be irregular, and piecewise procedures may be useful in these situations. In any case, careful examination of the data, especially by the plotting of individual curves, should be carried out before formal curve fitting is attempted.

4.1.2 Allometry and the Study of Shape

We have considered how the overall size of an individual or part of an individual changes with age. Not only does the size of an individual change with age, however, so too does his shape. At birth, for example, the head makes up about a quarter of the total length of the human body and this proportion decreases until adulthood when it makes up about one eighth of the total length. Such changes in the relative lengths of body segments are typical of most growing organisms and constitute what we perceive as change of shape. Broadly speaking, there have been two related but separate approaches to the study of shape changes with age. The first, pioneered by D'Arcy Thompson (1917) defines shape in terms of points or lines which correspond to particular positions on the individual—for example, on a two-dimensional frontal view of a child one point might be that corresponding to the end of the nose. The relative movement of these points or lines with increasing age is then the object of study, either pictorially as Thompson did with his transformation grid or by a suitable mathematical definition of shape in terms of the distances between the points and lines. The second approach seeks to find relationships between measurements which remain constant over time. This approach is motivated by considerations of growth dynamics and proceeds from assumptions concerning common processes underlying the measurements. The first major development using this latter approach was introduced by Huxley (1924, 1932) and is often referred to as Growth Allometry.

Growth allometry

The simplest form of relationship between measurements considers pairs of measurements at a time, and hence is also termed bivariate allometry. The most widely used equation relating pairs of measurements is known as the equation of simple allometry and may be derived as follows: suppose that growth is a process of self multiplication and hence that growth rate depends on the size of the organism. Suppose also that throughout the growth period the growth rate of a given organ slows down with increasing size and is actually proportional to size. Thus if x, y represent the sizes of two organs, for example head length and the length of the remainder of the body, we have

$$\frac{\mathrm{d}x}{\mathrm{d}t} = Ax \tag{4.6}$$

$$\frac{dy}{dt} = By \tag{4.7}$$

which leads, after eliminating time t, to

$$\log y = \alpha + \beta \log x \tag{4.8}$$

or

$$y = \alpha x^{\beta}$$

which is the usual form of the simple allometry equation. Figure 4.6 shows a plot of log (sitting height) on log (leg length) for a boy aged 4–13 years, and it can be seen that simple allometry is followed quite closely.

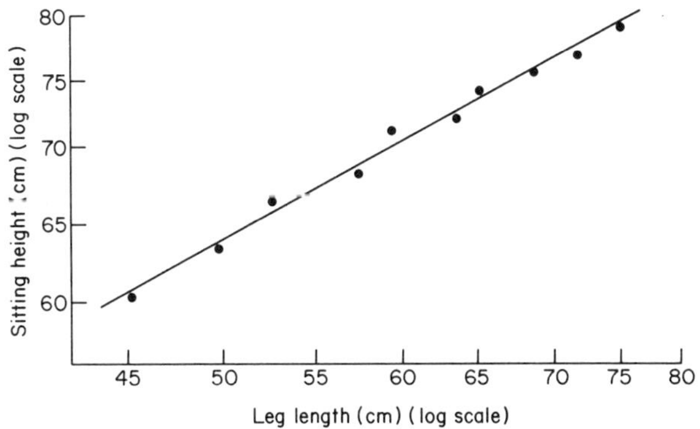

Fig. 4.6 Log (sitting height) (y) against log (leg length) (x) in a boy aged four years to 13 years, measured at yearly intervals. The fitted line is estimated as a functional relationship between the measurements assuming equal variances about the line. The equation is $y = 0{\cdot}905 + 0{\cdot}531x$.

We see immediately that equations (4.6) and (4.7) represent a different set of assumptions about growth rate than we made earlier in equation (4.1). Equation (4.7) for example leads to a growth curve of the form

$$y = Ke^{-bt}. \tag{4.9}$$

Although equations (4.6) and (4.7) are approximately equivalent to equation (4.1) for small values of y or values near to final size, for the reasons already discussed they are not adequate for the whole growth period—at least not for most animal growth. There are also other assumptions which can lead to equation (4.8). For example, if we have two curves in the form (4.3) then we also have simple allometry with

$$\beta = e^{(a_y - a_x)}$$

$$\alpha = \log k_y - \beta \log k_x$$

where the subscripts refer to the constants in (4.3) corresponding to the x and y measurements. For the logistic curve (4.2), however, simple allometry only holds if $\beta = 1$, i.e. the measurements are simply proportional to each other. This implies that the shape as measured by ratios of lengths is independent of size, a not very typical condition known as isometry. We see, therefore, that the equation of simple allometry is somewhat restrictive, although possibly useful as a characterization of shape in certain circumstances.

Various modifications of the simple allometry equation are possible. It seems that no more general equation than (4.8) fits every type of growth although particular extensions, such as including further terms like $(\log x)^2$ and $(\log x)^3$ in (4.8), fit certain types of growth. A review of various applications and extensions of the simple allometry equation is given by Gould (1966).

It has been pointed out by several authors (see Reeve and Huxley, 1945) that if two sections of an organ, for example a limb, both show simple allometry against say total body size, then their sum cannot exhibit simple allometry unless the parameter β is the same in each equation. Although the values of β will often be approximately equal, Reeve and Huxley do quote examples where they are very different.

Finally, on the subject of simple bivariate allometry, there is an important point to note on the method of actually estimating the values of α, β from a set of growth measurements. The logarithmic form of equation (4.8) looks like a simple linear regression equation of $\log y$ on $\log x$, but is not, in fact, since both x and y have equal status and are both subject to "error" or biological variation about the fitted line. It is a form of "functional relationship" and methods for estimation of α, β are given in, for example, Kendall and Stuart (1967, Chapter 29).

Multivariate allometry

Suppose we have several measurements k, say, among which we want to establish allometric relationships. A natural starting point is to consider all the possible pairs of relationships. For example, if there are three measurements x, y, z then the simple allometry equations are

$$\begin{aligned} \log y &= \beta_{yx} \log x + \log \alpha_{yx} \\ \log x &= \beta_{xz} \log z + \log \alpha_{xz} \\ \log z &= \beta_{zy} \log y + \log \alpha_{zy} \end{aligned} \tag{4.10}$$

where the subscripts refer to the measurements used. In fact, any two of these equations imply the third, or alternatively if $\log y$ is linearly related to $\log x$ and $\log x$ is linearly related to $\log z$ then $\log y$ is linearly related to log z. In this case the three measurements all lie on a straight line in three dimensions. As growth proceeds, its path can be followed along this line just as in Fig. 4.2 for two measurements. In practice, we would not expect a set of measurements to fall exactly on a straight line but they might have a small random variation about such a line. Hopkin (1966) discusses this situation in detail and provides statistical tests for examining how closely growth patterns conform to this simple allometric model.

At any given age, there will correspond a point on this straight line for an individual, and this moves along the line with age. Thus the distance along the line is related to the "size" of the individual and is a weighted average of the separate measurements. This definition of size can be written

$$S = \alpha_1 \log x_1 + \alpha_2 \log x_2 \ldots + \alpha_k \log x_k \tag{4.11}$$

where the weights α_i depend on the angle the line forms with each axis of measurement.

Where the variation about the line is random, the line can be estimated readily from a set of measurements as the best fitting line through the points at each age, and corresponds to the first "principal component". Some authors (e.g. Jolicoeur, 1963) have accordingly proposed to define size in general as being measured along such a best fitting straight line. Further non-random variation about this line can then be analysed for components of shape, in particular estimating further lines, orthogonal to the first, each representing a different component of shape. It should be noted, however, that this approach allows the data themselves to define size and defines shape components as independent of this, the results of which may not necessarily be very meaningful. A particular important case is where the α's equation in (4.11) are all equal, the isometry situation. Where true multivariate allometry exists then isometry implies that the individual's shape is constant.

In many circumstances it may be reasonable to define size simply as the sum or mean of the logarithms of all the measurements, namely,

$$S = \frac{1}{k}(\log x_1 + \log x_2 \ldots + \log x_k).$$

This has the useful property that any change of scale unit of an x_i changes S only by an overall constant amount, the same for each point. Also, since this holds for the isometry case when shape remains constant, it seems a natural choice for an *a priori* size variable. We may remove this size component from the measurements simply by calculating S for each point and forming the values

$$y_i = \log x_i - S$$

which can then be regarded as size adjusted measurements. These adjusted measurements can then be analysed for components of shape. Individual's shapes may then be compared, for example between ages or between different groups. An example where shape is used to discriminate between groups is given by Reyment and Banfield (1976). Corrucini (1972) gives an example of principal component analysis applied to such data. Burnaby (1966) and Gower (1976) give general procedures for adjusting measurements for a size factor prior to using the adjusted measurements for purposes of comparing or analysing the structures of groups of individuals.

If we wish to study change of shape with age it may be more convenient to have a single summary measure of shape rather than several components. One method of achieving this is to attempt to derive measures of shape at each age using a sample of individuals to estimate, say, the principal components of shape at each age. This enables us to form a weighted mean of the separate components. The problem with this approach seems to lie mainly in interpreting the separate components as measuring shape. Another approach would be to determine weights to be applied to the measurements by seeking those sets of weights which gave the highest correlations between composite shape variates across a series of ages. This procedure is described in the context of scaling techniques for categorical measurements by Healy and Goldstein (1976), and can also be used for continuous measurements.

Geometric methods for measuring shape

The alternative approach to estimating size and hence shape mentioned at the start of this section has been studied by Sneath (1967). This method of standardizing for size differs from the one described above in that it is done directly using diagrams of the individual at different ages, identifying corresponding points and then standardizing each one using the total sum of squared deviations of the set of points from the overall mean or centroid. This method seems to work fairly well with 2-dimensional representations and can in principle be extended to three dimensions. It does, however, require extensive and accurate measurement, and does not seem to have been widely used. It is also differently motivated to those we have already been discussing, in being concerned with the geometry of the 2-dimensional outline, whereas classical allometry is concerned with relationships between lengths of specified body components.

Having adjusted for size differences, we then consider shape in terms of shape change from one to another. If the diagrams are superimposed as closely as possible, the distances between corresponding points can be regarded as components of shape change and an overall measure obtained by, for example, calculating the total sum of such squared distances. Figure 4.7 shows two such superimposed diagrams of the same girl at ages $3\frac{1}{2}$ and 19 years taken from Goldstein and Johnston (1978).

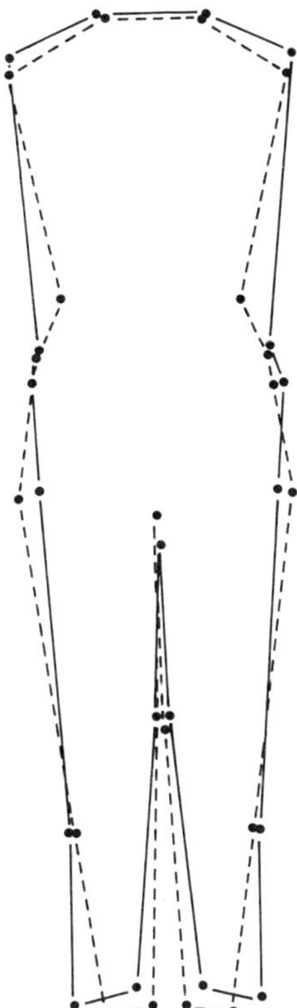

Fig. 4.7 Superimposed shape outlines of a girl at $3\frac{1}{2}$ years (continuous line) and 19 years (dashed line) after standardizing for size and orientation.

If a series of diagrams are available, then the shape change between the diagram at the youngest age and each subsequent diagram can be plotted against age. Alternatively we can see whether there is a simple relationship between the sets of points at two ages, which would then enable us to fully describe shape change in terms of a few parameters. Suppose, for example, that in both the x and y directions a straight line summarizes the change in coordinate values namely

$$x_2 = a + bx_1$$

$$y_2 = c + dy_1$$

where the subscripts 1, 2 refer to the x and y values at the first and second ages respectively. (We note that there is a fixed relationship between b and d which results from the method of size standardization.) Such relationships, established for the sample of points chosen, could then be used to transform the whole shape outline from one age to the next. In practice, however, such simple relationships are unusual and we would normally need to set up quite complex equations to describe fully the actual relationships. Once we have established a relationship, however, instead of using it to transform the diagram itself, we can transform the coordinate system to which it is referred. The idea of transforming such grids was suggested by D'Arcy Thompson (1917) and the above procedure may be regarded as a practical method of implementing Thompson's idea. Further details are given in Goldstein and Johnston (1978) who analyse the shape change of one child from 3 to 19 years.

By analogy with the previous discussion of size-standardized allometric measurements, we can similarly define shape change between two ages, as the average squared deviation of corresponding measurements, namely,

$$D = \frac{1}{k}\left((z_{21} - z_{11})^2 + (z_{22} - z_{12})^2 \ldots + (z_{2k} - z_{1k})^2\right)$$

where z_{21} is the first adjusted measurement at the second age, etc.

Before leaving the topic of shape measurement and change we make a brief reference to another system of shape measurement known as somatotyping (Sheldon *et al.*, 1954; Heath and Carter, 1967). This distinguishes three components of body shape, endomorphy, mesomorphy and ectomorphy and for each one defines seven points along a scale by reference to standard diagrams. In principle, therefore, every individual can be assigned a value on each component which together can be taken to represent shape. Such an approach seems not to have been adopted in growth studies of children and also raises some difficult problems of interpretation since the components are not independent of each other. A discussion and application of this method can be found in Tanner (1964).

Finally, it is worth noting that in studying shape we have been concerned with physical measurements. The question arises as to whether similar concepts can be applied to other measurements, such as behavioural ones, over time. The analogy here will be with allometry rather than a geometric measure of shape. In Chapter 3 we saw how mental and behavioural measurements are typically applicable only over short age spans. Thus, we have to deal with different measures at different ages and unless a direct calibration is possible, we have to resort to standardizing them separately with respect to the population of all individuals. This standardization, if it were designed to yield the same population distribution at each age, could be regarded as analogous to a size standardization, and mental test scores are often adjusted in this way. We could then search for constant relationships between measurements corresponding to simple allometry, the major differences here being that there seems to be no comparable simple theory relating pairs of mental or behavioural measurements as there is in the case of body components. We shall not therefore pursue this further.

4.2 Growth Curves Fitted to a Sample of Individuals

In Section 4.1 we discussed the estimation of the parameters of growth curves fitted to individuals. If each individual is regarded as belonging to a population of individuals, then just as we can consider the distribution of height at a given age over a population, so we can consider the distributions of the parameters of a growth curve. We may then study differences in a parameter or functions of this for different groups of individuals and relate them to further characteristics.

There is a further point which now arises. If we can summarize growth in terms of a few parameters for each individual, and if we know how these parameters are distributed in the population, then this should enable us to derive the distribution of the measurement itself at any given age. This particular line of enquiry, however, tends to lead to some difficult mathematics and has not been pursued very far. Krause *et al.* (1967) carry out the computations for 3-parameter logistic curves fitted to the weights of a sample of chickens. They make the assumption, however, that the parameters of the logistic are uncorrelated, whereas other investigators (e.g. Bock *et al.*, 1973) have found some quite high correlations. Krause *et al.* also find that the distributions they estimate in this way, whilst predicting the mean value quite well, do not seem to give very good estimates of the variability of the measurements at each age. If our purpose is to use growth curve summaries in this way, a more promising approach is by way of fitting polynomial curves, which is discussed below.

Before discussing this approach in detail we should mention a somewhat different approach suggested by Rao (1958). He pointed out that if we could find a transformation of the time scale, so that growth with respect to the new time scale was linear, then we could summarize growth simply in terms of the average rate of change. In one sense, this is the purpose of the various procedures for fitting individual growth curves that we have discussed, since once the form of the curve is known, this can be regarded as a transformation of the scale yielding an equation in which the measurement is proportional to the new transformed time parameter. If we have a population of individuals, an alternative is to define the time difference on the transformed scale between two occasions as equal to the difference in population means between the two occasions. Then, by definition, the population growth curve is linear and if we assume that for each individual, growth changes between occasions deviate only by a constant from the average population change, this transformation will lead to linear growth for each individual. Rao goes on to show how, if the population is divided into groups, we can use estimates from a sample to provide a valid test for equality of slopes between groups. He also proposes ways of testing the assumption of the same common transformation for each group.

The major difficulty with this approach seems to be the assumption that a *common* transformation exists. The empirical work on the fitting of individual physical growth curves seems to suggest that this is not the case. Furthermore, in many situations it would appear difficult to interpret differences in average growth rates using this kind of time scale which is defined solely in terms of the growth measurements themselves, and there seem to have been few applications of this technique.

4.2.1 Polynomial Growth Curves

A polynomial curve, as we have already seen, is a curve of the form

$$y_t = \alpha + \beta t + \gamma t^2 + \dots$$

and may be used to graduate growth. In this respect it has certain advantages over curves such as the logistic. First, it can be used to graduate any kind of growth curve more or less accurately by adding or deleting higher order terms. In fact, a polynomial with terms up to t^{p-1} can be made to pass exactly through p points. For the simpler polynomials a straightforward interpretation can be given to the coefficients. For example, for a straight line

$$y_t = \alpha + \beta t \tag{4.12}$$

the coefficient β measures the average growth rate for the individual. If a quadratic

$$y_t = \alpha + \beta t + \gamma t^2 \tag{4.13}$$

is fitted then the coefficient γ represents half the average growth acceleration. Furthermore, polynomials are not only simple to compute, it is also easier to work out the statistical distribution properties of the parameters when fitted to a sample of individuals than for curves such as the logistic. It is for this latter reason in particular that most of the important developments have taken place using polynomials.

Wishart (1938) seems to have been the first to use general polynomials to describe growth. He fitted a second order polynomial (4.13) to the growth in weight of different groups of pigs, and then separately analysed the linear and quadratic coefficients β and γ, looking for mean differences between the groups. Lacking in this approach, however, was a procedure for testing whether a low-order polynomial fitted the data well, or whether higher order terms were needed to provide a completely adequate description. Moreover, for the sample, the polynomial coefficients are correlated, so that, for example, a high average growth rate might be associated with a high acceleration, and we may find differences for both these parameters between groups of individuals. We might also be interested in whether the whole set of coefficients jointly defining the curve differs between groups.

These problems were tackled by Box (1950) who provided appropriate significance tests. Further developments along these lines were made in particular by Rao (1959, 1965) and Grizzle and Allen (1969) who showed how to provide confidence intervals. The following outline presents the basic ideas behind this approach, together with some other related approaches. Mathematical details can be found in appendices 4.1 and 4.2.

4.2.2 Multivariate Models

Let us first assume a simple situation in which we have a single sample of individuals each of whom is measured on each of p occasions. Later we shall generalize this to the case where the samples are drawn from different groups or populations and where not all individuals have measurements on all occasions. We recall that the essence of growth curve analysis is to make an adequate summary of the p measurements in terms of a small number of parameters. In the case of polynomial fitting the simplest summary will generally be in terms of a straight line defined by just an average value

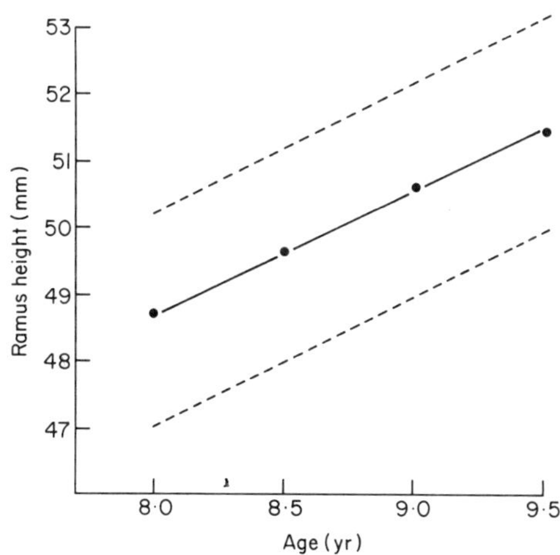

Fig. 4.8 Straight line fitted to mean Ramus height of 20 boys at four years, together with 95% parallel simultaneous confidence bands.

and a slope parameter. We consider first how to decide on an adequate summary of growth and we will assume this to be the same for all individuals.

Consider a sample of individuals each measured on four occasions. For simplicity we shall use some illustrative data which have also been analysed by other authors (see e.g. Elston and Grizzle, 1962). These consist of measurements of the height of the mandibular ramus bone in a sample of 20 boys taken at four half-yearly intervals from 8·0 years to 9·5 years (Table 4.1).

Table 4.1 Ramus Height of 20 Boys in mm

Individual	Age in years				Coefficient of fitted straight line $y = a + bt$	
	8·0	8·5	9·0	9·5	a	b
1	47·8	48·8	49·0	49·7	38·50	1·18
2	46·4	47·3	47·7	48·4	36·25	1·28
3	46·3	46·8	47·8	48·5	34·05	1·52
4	45·1	45·3	46·1	47·2	33·50	1·42
5	47·6	48·5	48·9	49·3	38·95	1·10
6	52·5	53·2	53·3	53·7	46·70	0·74
7	51·2	53·0	54·3	54·5	33·65	2·24
8	49·8	50·0	50·3	52·7	34·95	1·80
9	48·1	50·8	52·3	54·4	15·70	4·08
10	45·0	47·0	47·3	48·3	29·05	2·04
11	51·2	51·4	51·6	51·9	47·50	0·46
12	48·5	49·2	53·0	55·5	8·15	4·96
13	52·1	52·8	53·7	55·0	36·60	1·92
14	48·2	48·9	49·3	49·8	39·95	1·04
15	49·6	50·4	51·2	51·8	37·80	1·48
16	50·7	51·7	52·7	53·3	36·70	1·76
17	47·2	47·7	48·4	49·5	34·90	1·52
18	53·3	54·6	55·1	55·3	43·20	1·30
19	46·2	47·5	48·1	48·4	34·95	1·44
20	46·3	47·6	51·3	51·8	13·90	4·04
Mean	48·66	49·62	50·57	51·45	33·75	1·87
Standard deviation	2·52	2·54	2·63	2·73	10·20	1·16
Standard error	0·56	0·57	0·59	0·61	2·28	0·26

Figure 4.8 shows the means at each age, and it is clear that these lie very close to the fitted straight line. The mean value of the slope coefficient is about

seven times its standard error confirming that there is a trend with age, whereas when a quadratic equation involving the term t^2 is fitted to each individual, the mean value of the quadratic coefficient (c) is about half its standard error, and if we further fit a cubic term we find that the mean value of the cubic coefficient (d) is only about one tenth of its standard error. For completeness we may *jointly* test that both the quadratic and cubic terms are zero and this gives us an F value of 0·095 with 2 and 18 degrees of freedom. This is well below the conventional 5% significance level. A description of this test is given in Appendix 4.1. Following Wishart's (1938) example most investigators have used orthogonal polynomial regression coefficients rather than ordinary coefficients, but whilst this has certain computational advantages ordinary coefficients have a more convenient interpretation and we shall use them in what follows.

Table 4.2 Ramus Height Data: Correlations Between Coefficients of Cubic Polynomials

	c	d
a	–0·20	–0·05
b	0·23	0·87

Table 4.2 shows the correlation of the quadratic and cubic coefficients c and d with a and b when all four are fitted to each child.

We have already seen that c and d do not depart significantly from zero, but Table 4.2 shows that the linear and cubic coefficients have a high correlation. Hence it seems that whereas the average growth curve does not require a cubic term to describe it, within individuals the cubic coefficient is highly correlated with the linear, and can therefore be used to predict it. Hence, when estimating the linear coefficient, we may be able to use the predictive ability of the cubic coefficient to increase the *accuracy* with which the linear coefficient is estimated, and the method of doing this is outlined in Appendix 4.1.

Leech and Healy (1959) present an example where an estimate of the difference in growth rates between two groups is made more accurate by using associated initial differences between individuals in the two groups, for adjustment. Since the two groups were selected at random the expected mean initial difference was zero.

Although there are circumstances where the use of higher order coefficients can increase precision, some care is needed in their use since not all the coefficients in the equation are necessarily estimated more accurately. A further discussion of this point can be found in Grizzle and Allen (1969).

We now formalize this discussion by writing down the appropriate equations.

Suppose the measurement on the ith individual at age t is linearly related to t namely

$$y_{ti} = \alpha_i + \beta_i t + \varepsilon_{ti} \tag{4.14}$$

where α_i, β_i are the coefficients of the straight line for the ith individual and ε_{ti} is the residual variation about this line. If this model is an adequate representation then the ε_{ti} will be random variables with average value zero and independent of each other. They will, for example, include measurement errors and other unpredictable variations. Our present interest centres not so much on the individual's coefficients as on their average values, and their relationships with other variables. We can alternatively write (4.14) as

$$y_{ti} = \alpha + \beta t + \delta_{ti} \tag{4.15}$$

where the α, β are the expected values of α_i, β_i in the population and the δ_{ti} terms now incorporate the individual coefficients. In particular for an individual we have

$$\delta_{ti} = (\alpha_i - \alpha) + (\beta_i - \beta)t + \varepsilon_{ti}.$$

Hence, although the ε_{ti} are independent of each other for an individual, over the whole sample the values of δ_{ti} are not independent, since within an individual the term $(\alpha_i - \alpha) + (\beta_i - \beta)t$ is common to all the δ_{ti} values. For example, if α_i, β_i are high, that individual will generate a set of values all of which will tend to be higher than average. Thus the successive measurements are correlated, and although (4.15) looks like a straightforward simple regression equation, by virtue of the fact that the residual terms are not independent, the analysis becomes more complex.

In the above example there are four ages, and we can write down the covariances between the four residual terms ($t = 8{\cdot}0, 8{\cdot}5, 9{\cdot}0, 9{\cdot}5$) together with their variances as a matrix,

$$\Sigma = \begin{pmatrix} \sigma_{11} & \sigma_{12} & \sigma_{13} & \sigma_{14} \\ \sigma_{12} & \sigma_{22} & \sigma_{23} & \sigma_{24} \\ \sigma_{13} & \sigma_{23} & \sigma_{33} & \sigma_{34} \\ \sigma_{14} & \sigma_{24} & \sigma_{34} & \sigma_{44} \end{pmatrix}$$

where, e.g. σ_{12} is the covariance between the residuals from the first and second ages, etc., and the diagonal terms such as σ_{11} are the variances of the residuals attached to each age. If the measurements at each age had been made on different individuals then the covariance terms would be zero, and assuming equal variances, the analysis would simplify to the usual case of simple linear regression. In fact, the analysis also simplifies to the usual case if the off-diagonal terms are all equal and the diagonal terms are all equal. In the general case the non-zero correlation between residuals leads to a multivariate model, each individual having several (in the present case 4) dependent variables which are correlated among themselves.

The estimation of the parameters α, β together with Σ is described in Appendix 4.1. The calculation of standard errors and hence confidence intervals for α, β is described as well as a "simultaneous confidence band" for the estimated mean straight line. Figure 4.8 shows the estimated 95% confidence band for the fitted curve, $y = 33{\cdot}75 + 1{\cdot}876t$. The lines describing the confidence band may be thought of as those which will enclose the true population curve in 95% of samples (see Appendix 4.1).

In some situations we may be interested in intervals which include a given proportion of individuals' growth curves and this problem is discussed in Chapter 6 on growth standards.

We used the estimated coefficients for each individual in Table 4.1 to carry out some preliminary analyses. We can also use these coefficients to carry out the tests described above. In our example each individual has four measurements, and therefore for each individual we can fit a cubic polynomial which passes exactly through each measurement, say

$$y_t = a + bt + ct^2 + dt^3.$$

Hence we can effectively replace the four measurements by the four coefficients a, b, c, d, since knowledge of these enables us to calculate the measurement at each time point. By treating these four coefficients for each individual in just the same way as four measurements, we have an alternative method of studying the form of the polynomials, etc. We can, for example, test whether the mean values of c, d are zero, the test based on these coefficients being equivalent to that given for our original model (see Appendix 4.1).

The new model is (using a slightly different notation)

$$\begin{aligned} \beta_{0j} &= \beta_0 + \varepsilon_{0j} \\ \beta_{1j} &= \beta_1 + \varepsilon_{1j} \\ \beta_{2j} &= \beta_2 + \varepsilon_{2j} \\ \beta_{3j} &= \beta_3 + \varepsilon_{3j} \end{aligned} \tag{4.16}$$

where the subscripts 0, 1, 2, 3 refer to the constant, linear, quadratic and cubic coefficients, the j subscript refers to the individual and the ε_{0j}, ε_{1j}, etc., are the deviations of the individuals' coefficients from the mean coefficient. If the true curve is linear then β_2, β_3 are zero. An estimate of the correlation of ε_{0j} and ε_{1j} is the correlation between a and b and likewise for the higher order coefficients. We note, as before, that it is these correlations between the residuals which make the model a multivariate one. In the estimation of β_0, β_1 we can introduce β_2 as a covariate to increase precision as before and calculate a confidence band for the fitted line.

The use of the estimated individual's parameters gives the same results and in many ways is a more direct approach to the problem since the dependent variates are expressed simply in terms of the coefficients we wish to estimate. The reason for starting the discussion using the original measurements was to emphasize the link with ordinary polynomial regression.

4.2.3 Variable Occasion Times

One advantage of the above formulation is that it suggests a method for dealing with the case where not all measurements are made at the same occasions. This covers both the situation where for a given set of occasions some individuals may not have measurements on some occasions by accident or by design (see Chapter 2) and also the situation where some measurements are made near but not precisely at the "target" time. Thus in the above example, some measurements may be made at 8·05, 9·06 years, etc. One way of dealing with this latter situation is outlined in Chapter 7 and involves adjusting these measurements by fitting a low order polynomial to two or three surrounding measurements and interpolating an adjusted measurement at the required exact age. If we were to use all the available measurements to do this, then the coefficients of the polynomial used are the coefficients required in the above analysis. In the extreme case, measurements are not centred on occasions at all, but taken at quite different times on each individual. It is important to ensure that the time between measurements remains constant, however, otherwise seasonal or other biological, etc., effects may lead to different growth patterns for different individuals and to the residuals having different variances and covariances for different individuals (see Chapter 2). If the differences between occasions depart to a small extent only from equality, however, this procedure should give acceptable results, although little work seems to have been done on what amount of departure is allowable.

Where the same number of measurements are taken on each individual then we may fit the same order polynomial to each one and use the coefficients in our analysis as before. We should note that this procedure assumes that the same model is true over the total age range covered. It may be advisable to carry out a preliminary examination of this assumption by dividing individuals into groups (compatible with sufficient numbers in each group) covering different parts of the age range and comparing average coefficients between these groups.

4.2.4 Missing Data

When we do not have the same number of measurements on each individual, two approaches are possible. One approach develops a statistical model which explicitly takes account of the pattern of the data, and the other proceeds essentially by estimating the measurements which are missing by using the information from the measurements available.

The first approach is described by Kleinbaum (1973) and consists of dividing the data into sets of individuals, so that all the individuals within a set have measurements on the same set of occasions. Estimates of the growth curve coefficients and their standard errors, etc., are obtained by suitable combinations of estimates for each group. This approach will be convenient when there are a small number of different patterns. This may arise as a result of a mixed-longitudinal design of the type described in Chapter 2 with a few systematic patterns of measurement occasions. Often, however, measurements are absent on some occasions by chance, and especially where many occasions are involved, a large number of possible patterns may occur. The above approach may then become very cumbersome and can lead to troublesome situations where, for example, estimates of covariance matrices are obtained which are not "positive definite". An example of the application of Kleinbaum's method is given by Woolson *et al.* (1978).

We can avoid the difficulty caused by a large number of possible patterns by first calculating the polynomial regression coefficients for each individual. If there is a total of p occasions then there are at most $p - 1$ possible sets of polynomial coefficients and hence $p - 1$ possible patterns for analysis. The method described by Kleinbaum can be adapted to operate on the polynomial coefficients and details are given in Appendix 4.2. We note that we are making the tacit assumption here that the same growth curve form applies to all individuals. This may not always be the case, as when, for example, some individuals have measurements missing at one end of the scale and some at the other end, leading to different forms of growth curve. A preliminary inspection of the data will often provide information about such problems.

The second approach is to manipulate the data in order to fill in those occasions which are missing with estimated values, so that the analysis can then proceed in the usual way (with some modifications). A general method of tackling this problem for multivariate data is to begin by selecting those individuals where all the measurements have been made. Among the remaining individuals we identify the patterns for which some measurements are missing and some present. For each missing measurement we use a group of individuals with complete data to calculate the best prediction of the missing measurement. The prediction (using, say, multiple regression) is then applied to the measurements from the incomplete set and the missing measurement replaced by its predicted value. From these data, new estimates of the means and covariances of the measurements can be obtained which are then used again to predict the missing values. The process is repeated until stable predicted values of the missing data are obtained. A description of this procedure with certain modifications can be found in Beale and Little (1975) and Dempster *et al.* (1977). With growth data this procedure can be simplified as in the first approach by first calculating polynomial regression coefficients. An individual with some missing measurements will therefore lose some of the highest order coefficients, whilst all individuals will provide estimates for the low order coefficients.

We may now carry out the procedure for filling in the missing high order coefficients using the sets of low order coefficients as predictors. When only a few measurements at most are missing from any one individual this considerably reduces the amount of computation required since only a few different patterns exist. In many practical situations, only a very few individuals may have more than one or two measurements missing and the omission of these individuals from the analysis may considerably cut computational complexity without losing much information.

4.2.5 Successive Difference Methods

Suppose growth is linear. For any individual measured at p occasions we can obtain $p - 1$ estimates of his growth rate simply by calculating the rate of change of the measurement from one occasion to the next, i.e.

$$v_{i+1} = (y_{i+1} - y_i)/(t_{i+1} - t_i),\ i = 1, \ldots, p - 1.$$

Where y_i, t_i are the values of the measurement and time at occasion i. The v_i are known as first order divided differences. Since growth is assumed to be linear, the expected values of the v_i will each be equal to the growth rate. Hence the expected values of the differences between the first order divided differences, namely the second order differences,

$$v_{i+1} - v_i, \quad i = 1, \ldots, p - 2$$

will be zero. Thus we can construct a test of whether the growth curve is linear by using these second order differences. Each individual has a set of $p - 2$ such differences, and using the n sets of differences we can readily test the hypothesis that their expected values are zero. In fact the test is equivalent to that given in Appendix 4.1 for testing the order of the polynomial fitted to the measurements. Hills (1968) gives a full description of this approach. Although there is no particular computational advantage, this does allow us to explore the structure of the data in some detail. We can, for example, average second order differences at different parts of the time scale to see whether a straight line might be adequate in one part and not in another. Thus a pattern of growth which took place in two parts, one linear and one quadratic could be detected in this way, whereas fitting an an average curve to the whole period would not show this up. For the mandibular ramus bone data the average second order differences with their standard errors are as follows. The first difference is estimated from the first three occasions and the second from the last three.

Difference	Mean	Standard Error
1	–0·025	0·25
2	–0·065	0·23

Both of these mean values are well below their standard errors, and as we saw earlier, the combined test for the second order differences equal to zero gives a non-significant result.

4.2.6 Group Differences

So far we have been concerned with estimating growth curves for a single homogenous group of individuals. We will often, however, wish to make comparisons between groups of individuals in terms of the coefficients of the polynomials which fit the individuals in the different groups. Suppose, for example, that a quadratic curve was fitted to each of m groups – we might then wish to say whether the mean quadratic coefficients were equal or whether the linear and quadratic coefficients jointly were equal.

Using a similar notation to (4.15) for linear growth, we can write the value of a measurement on the ith individual at age t in the jth group as

$$y_{tij} = \alpha_j + \beta_j t + \delta_{tij}. \tag{4.17}$$

The analysis, as outlined in Appendix 4.1, will normally proceed in two parts. The first estimates the order of the polynomial curves within each group. It is assumed initially that the same order polynomial is present in each group. In the second part comparisons are made between the p polynomial coefficients either singly or jointly. For each group we consider the set of dependent variables as the polynomial coefficients and their mean values are compared between groups.

We can readily extend the simple comparison of groups to more complicated designs where individuals may be cross-classified, e.g. by social class and sex, and continuous variables, or covariates, may also be introduced into the design, e.g. height of a child's parents. As we remarked before, a special case of such covariates is the use of higher order coefficients with average value zero but non-zero correlations with the low-order coefficients.

The following simple example compares growth curves for three groups of pre-adolescent girls classified according to height of their mothers. Five measurements are made at yearly occasions from ages six to 10. (N.B. the measurements are given at exact years of age, some having been previously adjusted to these.)

Table 4.3 gives the raw data and the means for three groups which are plotted in Fig. 4.9, using data from the London Growth Study (Tanner *et al.*, 1976).

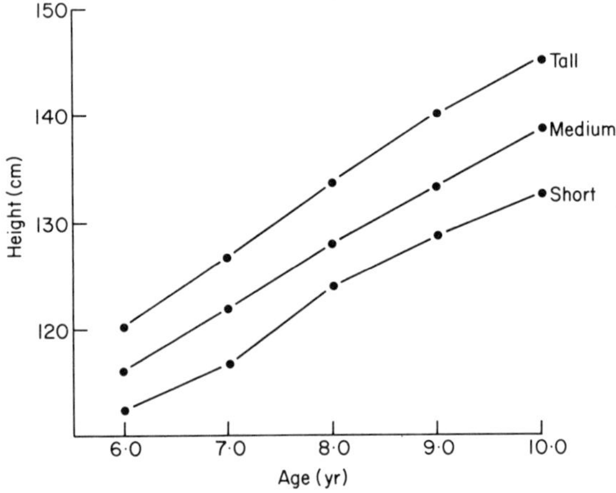

Fig. 4.9 Mean height of six to 10-year-old girls for short (< 155 cm) medium (155 – 164 cm) and tall (> 164 cm) mothers.

From the figure it seems that straight lines might fit the data. A test for zero higher order coefficients, however, gives a χ^2 on nine degrees of freedom of 27.3 which is significant at the 1% level.

Thus we need at least a quadratic curve over this age range to describe height growth. For simplicity, however, we shall restrict this analysis initially to the straight line fit so that the focus of interest is on *average* rate of change over this period. Table 4.4 gives the fitted slope and intercept for the three groups without using any higher order terms as covariates.

A test for slope differences between groups gives a χ^2 on two degrees of freedom of 9·5 which is just significant at the 1% level. A joint test for intercept

Table 4.3 Height (cm) of Girls According to Height of their Mothers

Individual	6·0	7·0	Age 8·0	9·0	10·0
Short mother					
1	111·0	116·4	121·7	126·3	130·5
2	110·0	115·8	121·5	126·6	131·4
3	113·7	119·7	125·3	130·1	136·0
4	114·0	118·9	124·6	129·1	134·0
5	114·5	112·0	126·4	131·2	135·0
6	112·0	117·3	124·4	129·2	135·2
Mean	112·5	116·7	124·0	128·7	133·7
Medium mother					
7	116·0	122·0	126·6	132·6	137·6
8	117·6	123·2	129·3	134·5	138·9
9	121·0	127·3	134·5	139·9	145·4
10	114·5	119·0	124·0	130·0	135·1
11	117·4	123·2	129·5	134·5	140·0
12	113·7	119·7	125·3	130·1	135·9
13	113·6	119·1	124·8	130·8	136·3
Mean	116·2	121·9	127·9	133·2	138·5
Tall mother					
14	120·4	125·0	132·0	136·6	140·7
15	120·2	128·5	134·6	141·0	146·5
16	118·9	125·6	132·1	139·1	144·0
17	120·7	126·7	133·8	140·7	146·0
18	121·0	128·1	134·3	140·3	144·0
19	115·9	121·3	127·4	135·1	141·1
20	125·1	131·8	141·3	146·8	152·3
Mean	120·3	126·7	133·6	139·9	144·9

and slope differences between groups gives a χ^2 on four degrees of freedom of 20·6 which is significant at the 0·1% level. We conclude therefore that not only are there average differences for the maternal height groups but also that the girls from the taller groups grow more rapidly over this age range.

Fitting quadratic terms a test for equality gives a χ^2 on two degrees of freedom of 3·3 so that we may conclude that there is no evidence to suggest that the acceleration of growth is different for the three groups. A test for the average quadratic parameter equal to zero gives a χ^2 on 1 degree of freedom equal to 7·5 which is significant at the 1% level.

Table 4.4 Estimate of Mean Intercept and Slope Parameters for Data in Table 4.3

	Short mothers	Medium mothers	Tall mothers
a	79·6	83·0	83·1
b	5·44	5·57	6·25

4.2.7 Non-parametric Methods

The above methods of analysis assume that the measurements have normal or near normal distributions or can be transformed to normality. In some situations, however, this cannot be done easily and it is natural to turn to non-parametric techniques which rely only on the relative ordering of values. We might, for example, have a five point rating scale such as a puberty rating or an attainment rating, and wish to compare rates of progress for different groups of individuals over a set of occasions. We might begin by defining a measure of rate of progress by, say, calculating the regression coefficient of the rating (with suitably scored categories) on time, which together with the average rating would be used to characterise the individual. We could, of course, calculate higher order coefficients if we wished and thought they were meaningful. We note that the degree of polynomial now has to be determined at the outset and there appears to be no procedure analogous to that given above for testing the degree of the polynomial which fits the data.

The procedure used to compare groups is as follows. First we consider the univariate case where we are studying just one individual parameter, say the average. We can use a rank-sum test which consists of ranking all individuals in order of magnitude, giving the smallest value rank = 1, etc., calculating the average scores of these ranks separately for each group and then forming a test statistic based on the differences between the mean ranks (see e.g. Siegel, 1956). For multiple parameters, e.g. mean and rate of change, we can extend the procedure to a multivariate rank-sum test, with parameters being compared jointly between groups. Ghosh *et al.* (1973) discuss this analysis and show how multiple ways of classification of individuals can also be introduced. An interpretation of the analysis can be made in terms of differences between the mean rank scores of the different groups for each parameter. The idea behind this approach is thus relatively straightforward and does not introduce any essentially new ideas. We confine ourselves here to a simple illustration using some artificial data on teachers' gradings of children's reading ability on a 10-point scale. There are two groups of individuals from two different social groups, each measured on three occasions a year apart.

Table 4.5 Reading Ability Gradings (Hypothetical Data)

Child	Occasion 1	2	3	Mean Score	Rate of Change
Social Group A					
1	1	2	4	2·3 (2·5)	1·5 (5)
2	2	2	3	2·3 (2·5)	0·5 (1)
3	1	3	5	3·0 (5)	2·0 (11)
4	1	3	4	2·7 (4)	1·5 (5)
5	3	5	7	5·0 (13·5)	2·0 (11)
6	3	3	6	4·0 (9·5)	1·5 (5)
7	2	5	6	4·3 (11)	2·0 (11)
8	2	3	5	3·3 (7)	1·5 (5)
9	1	1	4	2·0 (1)	1·5 (5)
10	2	3	7	4·0 (9·5)	2·5 (16·5)
11	3	6	8	5·7 (15)	2·5 (16·5)
Mean Rank				(7·32)	(8·18)
Social Group B					
1	4	4	6	4·7 (12)	1·0 (2)
2	3	7	9	6·3 (17)	3·0 (19·5)
3	2	2	6	3·3 (17)	2·0 (11)
4	1	4	5	3·3 (7)	2·0 (11)
5	3	4	8	5·0 (13·5)	2·5 (16.5)
6	5	7	9	7·0 (19)	2·0 (11)
7	4	6	10	6·7 (18)	3·0 (19·5)
8	4	5	9	6·0 (16)	2·5 (16·5)
9	6	8	10	8·0 (20)	2·0 (11)
Mean Rank				(14·39)	(13·11)

Figures in brackets are rank positions of individuals' calculated mean and slope parameters, allowing for ties.

We first carry out a univariate rank-sum test and obtain a highly significant difference (using an exact test) between the mean ranks. The difference between the means of the ranks of the rates of change almost reaches the 5% significance level. We now compare the mean and rate of change jointly between groups and obtain a χ^2 of 7·21 on two degrees of freedom which is significant at the 5% level, but we would need to interpret this significance level with a little caution in view of the small sample size. On the strength of these results, we could conclude that the ability profiles were different for the social groups, the differences being more marked for the mean score than the rate of change. The Spearman rank correlation coefficient between the mean score and rate of change is 0·66, which is relatively high, suggesting that there is a strong tendency for children with overall high ratings to progress faster.

Where there are many ties, that is the number of distinct values taken by any variable is small, we could use the approach described in Chapter 5 dealing with the analysis of categorical data. Zerbe and Walker (1977) study a randomization test for comparing groups when the order of the polynomial is not necessarily the same in each group.

4.2.8 Further Considerations on Growth Curve Models

The above discussion has dealt with general procedures for estimating and testing the coefficients of polynomial growth curves. The parametric theory makes use of the assumption that the measurements have a normal distribution, but allows a completely general structure for their variances and covariances. That is, they may have different variances and correlations. In some circumstances, however, we may have extra information about the relationships among these variances and correlations which we would want to utilise in order to obtain more powerful tests, smaller standard errors and narrower confidence intervals. One such pattern is considered in detail by Rao (1965) and given in Appendix 4.1. Detailed discussions of various possibilities are also given by Khatri (1973) and Morrison (1972). Joreskog (1970) also gives a discussion in the context of a general method for the analysis of covariance structures. One point worth mentioning again is that when all the intercorrelations are equal, we can carry out at valid univariate analysis without resorting to the multivariate model (Scheffe, 1959). In practice this situation seems to be rare, since occasions further apart in time will tend to be less highly correlated. If we use a univariate analysis when this assumption is false we may obtain very biased estimates and tests. Elston and Grizzle (1962) give an example using the mandibular ramus bone data we used in our first example and show that the univariate analysis gives confidence intervals which are too narrow.

Although there seems to be no particular advantage in attempting to use a univariate model, it is possible to carry out approximately valid tests by appropriate adjustments to the degrees of freedom of certain test statistics. Huynh and Feldt (1976) give a discussion of this and an application to growth curves can be found in Greenhouse and Geisser (1959).

The normality assumption may also be important. For the data given in the examples here and in general for measurements such as height, the assumption is well enough satisfied. Where the distributions depart markedly from normality, then we should consider a transformation which produces approximate normality, or carry out a non-parametric analysis. A plot of the data on normal probability paper will be a useful preliminary to any analysis.

We have also assumed that the samples of individuals have been simple random ones from a well defined population. Where a complex sampling scheme has been used, some modifications may be necessary. For a stratified sample, we may be able to allow for strata differences in the between-individuals component of the model. Where clustering has been used, the problems tend to be more difficult and the reader is referred to the discussion in Chapter 2.

Appendix 4.1

Estimation and testing for polynomial growth curves

The following outline is based mainly on the exposition by Grizzle and Allen (1969) to which reference can be made for a more detailed discussion.

For each individual, we suppose that a characteristic is measured at p time points. For an individual we can write the relationship between the measurement vector and time as follows

$$X = B\xi + e \tag{4.1.1}$$

where X is the $(p \times 1)$ vector of measurements; B is a $(p \times q)$ design matrix; ξ is a $(q \times 1)$ vector of unknown coefficients; e is a $(p \times 1)$ vector of residuals.

The degree of the polynomial fitted to the data is therefore $q - 1$. For example, if $p = 4$ and $q = 2$ we have

$$B = \begin{bmatrix} 1 & t_1 \\ 1 & t_2 \\ 1 & t_3 \\ 1 & t_4 \end{bmatrix}$$

$$\xi = \begin{bmatrix} \beta_0 \\ \beta_1 \end{bmatrix}$$

giving the set of equations (4.2).

We may further classify individuals according to their characteristics, such as sex and social class. We can incorporate this into the model by defining a further design matrix A and writing for the whole sample

$$X = B\,\xi\,A + E. \tag{4.1.2}$$

X is now $p \times N$ where N is the total number of individuals; B is as before, the within-individual design matrix; ξ is $q \times r$; A is $r \times N$, the between-individual design matrix; E has columns which are independently, normally distributed with zero mean and covariance matrix Σ; r is the number of independent variables used to classify individuals.

For example, if individuals are classified into three groups A may be written

$$A = \begin{bmatrix} 1 & 1 \dots 1 & 0 & 0 \dots 0 & 0 & 0 \dots 0 \\ 0 & 0 \dots 0 & 1 & 1 \dots 1 & 0 & 0 \dots 0 \\ 0 & 0 \dots 0 & 0 & 0 \dots 0 & 1 & 1 \dots 1 \end{bmatrix}$$

The same polynomial structure is assumed for each element of the between individuals design, but some of the coefficients in ξ may be zero, implying different order polynomials for different groups of individuals. Also A can be completely general, incorporating continuous as well as discrete or categorized variables. Now write $B_1 = B(B^T B)^{-1}$;

B_2 as a $p \times (p - q)$ matrix such that $B_2^T B = 0$.

If we have used orthogonal polynomials, then B_2 may be chosen with columns which are the higher order polynomial terms not included in B. Alternatively we may use the matrix which produces the qth order differences (or divided differences) as described in Section 4.2.5. In the above example of linear growth with equally spaced time intervals this gives

$$B_2^T = \begin{Bmatrix} 1 & -2 & 1 & 0 \\ 0 & 1 & -2 & 1 \end{Bmatrix}.$$

Transform X as follows

$$Y_1 = B_1^T X$$
$$Y_2 = B_2^T X.$$

Hence

$$Y_1 = (B^T B)^{-1} B^T X$$

and is thus the $q \times n$ matrix of individual polynomial coefficients up to order $(q - 1)$, and

$$E(Y_1) = \xi A$$

also $$E(Y_2) = 0.$$

The complete model can be written

$$\begin{Bmatrix} Y_1 \\ Y_2 \end{Bmatrix} = \begin{Bmatrix} \xi \\ \beta \end{Bmatrix} A + E_1 \tag{4.1.3}$$

where $\beta = 0$ and the columns of E_1 are independently normally distributed with mean vector zero and covariance matrix

$$\Sigma^* = \begin{Bmatrix} B_1^{\mathrm{T}} \Sigma B_1 & B_1^{\mathrm{T}} \Sigma B_2 \\ B_2^{\mathrm{T}} \Sigma B_1 & B_2^{\mathrm{T}} \Sigma B_2 \end{Bmatrix}.$$

The first stage in examining the model is usually to ascertain the order of polynomial growth curve. This can be done simply as follows.

Ignoring the within individual design we can write

$$E(X) = \eta A, \tag{4.1.4}$$

where η now refers to the mean values at each occasion.

The hypothesis

$$H_0 : B_2^{\mathrm{T}} \eta I = 0 \tag{4.1.5}$$

is therefore a test that polynomial coefficients of order q to p are zero. The hypothesis matrix is

$$P_1 = B_2^{\mathrm{T}} X A^{\mathrm{T}} (AA^{\mathrm{T}})^{-1} A X^{\mathrm{T}} B_2$$

and the error matrix is

$$Q_1 = B_2^{\mathrm{T}} X \{I\text{-}A^{\mathrm{T}} (AA^{\mathrm{T}})^{-1} A\} X^{\mathrm{T}} B_2.$$

The likelihood ratio test criterion can be written

$$U = -m \log \prod_{i=1}^{p-q} \frac{1}{1+\lambda_i}$$

where $$m = \{N - r - \tfrac{1}{2}(p - q - r + 1)\}$$

and λ_i are the latent roots of the $(p - q) \times (p - q)$ matrix $P_1 Q_1^{-1}$. Under the null hypothesis U has asymptotically a χ^2 distribution with $r(p - q)$ degrees of freedom.

Having established the order of the polynomial growth curve we can test further hypotheses about the parameters in ξ. This may be done using (4.1.3) as follows. The simplest situation is when the vectors Y_2 provide no information about Y_1 and are ignored and assigned solely to error. The model is then

$$E(Y_1) = \xi A$$

and the maximum likelihood estimate of ξ is $\hat{\xi} = Y_1 A^{\mathrm{T}} (AA^{\mathrm{T}})^{-1}$ and we may test hypotheses of the form

$$H_0 : C \xi V = 0 \tag{4.1.6}$$

where C is $(c \times q)$ and is concerned with the within-individual parameters and V is $(r \times v)$ and is concerned with the between-individual parameters.

The hypothesis matrix is $P_1 = (C \hat{\xi} V) \{V^{\mathrm{T}} (AA^{\mathrm{T}})^{-1} V\}^{-1} (C \hat{\xi} V)^{\mathrm{T}}$.

The error matrix is $Q_1 = C Y_1 \{I - A^{\mathrm{T}} (AA^{\mathrm{T}})^{-1} A\} Y_1^{\mathrm{T}} C^{\mathrm{T}}$.

In the above example we may wish to test whether the linear coefficient in group 3 is zero. We take

$$C = (0 \quad 1)$$

$$V = \begin{pmatrix} 0 \\ 0 \\ 1 \end{pmatrix}.$$

The test criterion is derived in a similar manner to before with $m = N - r - \frac{1}{2}(c - v + 1)$ and degrees of freedom $= cv$.

In some situations, as discussed in the text, the vectors in Y_2 may provide information on Y_1 and can be used as covariates, added to the design matrix A. We can then write (4.1.3) as

$$Y_1 = \left\{\xi : n\right\} \left\{\begin{matrix} A \\ \ddot{} \\ Y_2 \end{matrix}\right\} + E_2$$

or simply

$$Y_1 = \gamma D + E_2 \tag{4.1.7}$$

with covariance matrix $\Sigma_1 = (B^T \Sigma^{-1} B)^{-1}$.

The maximum likelihood estimate of γ is

$$\hat{\gamma} = Y_1 D^T (DD^T)^{-1}$$

and the first r columns give the required estimate of ξ.

Explicitly the estimate is given by

$$\hat{\xi} = (B^T S^{-1} B)^{-1} B^T S^{-1} X A^T (AA^T)^{-1}$$

where $S = X\{I\text{-}A^T (AA^T)^{-1} A\} X^T$ is an estimate of the residual dispersion matrix. A consistent estimate of the variance of $\hat{\gamma}$ is given by $(DD^T)^{-1} \otimes \Sigma_1$ where $\otimes$ is the Kronecker product. An unbiased estimate of Σ_1 is given by

$$\hat{\Sigma}_1 = \frac{(B^T S^{-1} B)^{-1}}{n - r - (p - q)}.$$

Hypotheses of the form (4.1.6) applied to the model (4.1.7) can readily be tested. Thus when covariates are used, the hypothesis (4.1.7) becomes

$$H_0 = C\left\{\xi : \eta\right\} \left\{\begin{matrix} V \\ \ddot{} \\ 0 \end{matrix}\right\} = C \xi V = 0.$$

Where k covariates are used, m is reduced by k. We may also test hypotheses about η, and in particular whether a subset of these coefficients are zero implying that the corresponding covariates add no information.

Confidence Intervals

We can select a subset of the parameters of ξ, by using $C\,\xi\,V$ with suitable values of C and V.

Any linear function of these parameters can be defined using vectors a, b in the expression

$$a^{\mathrm{T}} C\,\xi\,V b. \tag{4.1.8}$$

Thus, if we wish to define the predicted mean value at the fourth time point on the straight line fitted to the third group in the example (Table 4.3), we should take

$$C = \begin{pmatrix} 1 & 0 \\ 0 & 1 \end{pmatrix};\ V = \begin{pmatrix} 0 \\ 0 \\ 1 \end{pmatrix};\ a = (1 \quad t_4);\ b = (1).$$

Simultaneous confidence intervals for all a, b in expression (4.1.8) are given by

$$a^{\mathrm{T}} C\hat{\xi} V b \ \pm \{\alpha(a^{\mathrm{T}} Q_2 a)\,(b^{\mathrm{T}} V^{\mathrm{T}} R V b)\}^{\frac{1}{2}}$$

where $Q_2 = C(B^{\mathrm{T}} S^{-1} B)^{-1} C^{\mathrm{T}}$,
and R is the $r \times r$ minor of DD^{T} corresponding to the elements of ξ in (4.1.5).
$\lambda_\alpha = \dfrac{u_\alpha}{1 - u_\alpha}$ where u_α is the αth percentage point of the largest root test criterion with the usual notation for degrees of freedom (Pillai, 1967).
$S = \min(c, v)$, $m = (|v - c| - 1)/2$,
$n = \{N - r - (p - q) - c - 1\}/2$.

Since these intervals provide simultaneous bounds for all a, b we can give confidence limits for the whole time scale. An example is shown in Fig. (4.8).

Covariance Structures

Rao (1965) proposed a particular structure for the covariance matrix of measurements Σ. This had the form

$$B\,\Gamma\,B^{\mathrm{T}} + I\sigma^2.$$

This can be motivated by considering each individual's polynomial coefficients to be random variables with dispersion matrix Γ, together with independent errors with variance σ^2. Other assumptions about the structure of Σ can be made and discussions are given by Khatri (1973), Joreskog (1970) and Morrison (1972). If we can assume a particular covariance structure for the model, this should lead to more powerful significance tests and shorter confidence intervals. It seems, however, that large amounts of data would be necessary to discriminate between models. In some applications there may be external reasons for supposing a particular covariance structure to hold and this could be incorporated into the analysis. It should also be noted that, if the covariance matrix were known, or a good independent estimate of it were available, we could use a generalized least squares estimator for the parameters (Lee, 1974). Fuller and Battese (1973) show how to construct such an estimator for the case where the residual error is assumed to be the sum of a random error for each individual and a random error associated with each measuring occasion.

Chakravorti (1975) has considered the model where the measurements have different covariance matrices in different cells of the between-individuals design. Bayesian methods which specify prior distributions for the parameters of the model have also been used, and the interested reader is referred to Fearn (1975, 1977).

Appendix 4.2

Estimation and Testing in the Growth Curve Model with Missing Data

We consider the following model where the dependent variates are the individual polynomial regression coefficients with up to $(p - 1)$ possible patterns deriving from the number of coefficients which can be estimated.

Firstly, we can determine the order of polynomial to be fitted by a straightforward application of Kleinbaum's method (1973) in which the dependent variates are the polynomial coefficients and the within-individual design matrix is the unit matrix. Alternatively, if we wish to test, say, the two highest order polynomial coefficients, we can carry out the test based on the hypothesis (4.1.5) separately in the two groups of individuals, those missing no occasions and those missing one occasion. In the latter group, of course, the test will only involve the second highest order coefficient. We then sum the χ^2 statistics and degrees of freedom for each test and judge the significance of the combined statistic.

We now consider the model 4.1.7. We shall assume that all individuals have at least q measurements so that the missing data belong to the matrix Y_2. Suppose there are $j \leqslant p - q$ patterns of missing data, numbered $1 \ldots j$ so that the first group has no missing occasions, the second has one occasion missing, etc. For the ith group we can write

$$E(Y_{1i}) = \eta \, C_i \, D$$

where C_i consists of 1's and 0's defining the missing elements of Y_2 belonging to D and Y_{1i} is the matrix of values of the ith group of Y_1. For example, for the second group where the highest order coefficient is missing we can write

$$C_2 = \begin{pmatrix} 1 & \cdots & 0 \\ \vdots & \ddots & \vdots \\ \vdots & & 1 \\ 0 & & 0 \end{pmatrix}.$$

We now define the vector $y = (y_1^{\mathrm{T}} y_2^{\mathrm{T}} \ldots y_j^{\mathrm{T}})^{\mathrm{T}}$ where y_i is the $(n_i q \times 1)$ vector formed by placing the columns of Y_1 for ith group underneath each other. The model can now be written

$$E(y) = F\eta \tag{4.2.1}$$

where η is arranged as a column vector by placing the columns underneath each other

and
$$F = \begin{pmatrix} D^{\mathrm{T}} C_1 \otimes K_{n_j} \\ \vdots \\ D^{\mathrm{T}} C_j \otimes K_{n_j} \end{pmatrix}$$

where K_{n_j} is the unit vector of length n_j,

$$\text{and Variance } (y) = \Omega = \begin{pmatrix} C_1^{\mathrm{T}} \Sigma_1 C_1 \otimes I_{n1} & & \\ & \ddots & \\ & & C_j^{\mathrm{T}} \Sigma_1 C_j \otimes I_{n_j} \end{pmatrix}.$$

(4.2.1) is in the form of a univariate weighted LS model and a BAN estimate of η is

$$\hat{\eta} = (F^{\mathrm{T}} \hat{\Omega}^{-1} F)^{-1} F^{\mathrm{T}} \Omega^{-1} y$$

with covariance matrix $\hat{V} = (F^{\mathrm{T}} \hat{\Omega}^{-1} F)^{-1}$.
After manipulation we obtain

$$\hat{\eta} = \left\{ \sum_1^j C_i^{\mathrm{T}} D (C_i^{\mathrm{T}} \Sigma_1 C_i)^{-1} D C_i \otimes K_{n_i}^{\mathrm{T}} K_{n_i} \right\}^{-1}$$

$$\times \sum_1^j \left\{ C_i^{\mathrm{T}} D (C_i^{\mathrm{T}} \hat{\Sigma}_1 C_i)^{-1} \otimes K_{n_i}^{\mathrm{T}} \right\} y_i$$

$$\hat{V} = \left\{ \sum_1^j C_i^{\mathrm{T}} D (C_i^{\mathrm{T}} \hat{\Sigma}_1 C_i)^{-1} D^{\mathrm{T}} C_i \otimes K_{n_i}^{\mathrm{T}} K_{n_i} \right\}^{-1}.$$

We need a consistent estimator of $\hat{\Sigma}_1$, and this can be obtained by forming separate estimates for each value of i and pooling the estimates.

To test the hypothesis

$$H^{\mathrm{T}} \eta = 0 \tag{4.2.2}$$

where H is $(rq \times w)$ and has rank w. We form the statistic

$$W = (H^{\mathrm{T}}\hat{\eta})^{\mathrm{T}} \{H^{\mathrm{T}} \hat{V} H\}^{-1} H^{\mathrm{T}}\eta$$

which under the null hypothesis is distributed as χ^2 with w degrees of freedom.

This hypothesis is derived from the hypothesis

$$C \eta V = 0$$

by setting $H^{\mathrm{T}} = V^{\mathrm{T}} \otimes C$.

References

Beale, E. M. and Little, R. J. A. (1975). Missing values in multivariate analysis, *Journal of the Royal Statistical Society*, B **37**, 129–145.

Boas, F. (1892). The growth of children, *Science*, **20**, 351–352.

Bock, R. D. and Thissen, D. M. (1976). Fitting multi-component models for growth in stature; unpublished mimeo, University of Chicago.

Bock, R. D., Wainer, H., Peterson, A., Thissen, D., Murray, J. and Roche, A. (1973). A parametrization for individual human growth curves, *Human Biology*, **45**, 63–80.

Box, G. E. P. (1950). Problems in the analysis of growth and wear curves, *Biometrics*, **6**, 362–389.

Burnaby, T. P. (1966). Growth invariant discriminant functions and generalized distances, *Biometrics*, **22**, 96–110.

Burt, C. B. (1937). "The Backward Child" University of London Press.

Chakravorti, S. R. (1975). Maximum likelihood estimation for the growth curve model with unequal disperson matrices, *Communications in Statistics*, **4**, 69–85.

Coruccini, R. S. (1972). Allometry Correction in taxometrics, *Systematic Zoology*, **21**, 375–385.

Deming, J. (1957). Application of the Gompertz curve to the observed pattern of growth in length of 48 individual boys and girls during the adolescent cycle of growth, *Human Biology*, **29**, 83–122.

Dempster, A., Laird, N. M. and Rubin, D. B. (1977). Maximum likelihood from incomplete data via the EM algorithm, *Journal of the Royal Statistical Society*, B **39**, 1–38.

Elston, R. S. and Grizzle, J. E. (1962). Estimation of time response curves and their confidence bands, *Biometrics*, **18**, 148–159.

Fearn, T. (1975). A Bayesian approach to growth curves, *Biometrika*, **62**, 89–100.

Fearn, T. (1977). A two-stage model for growth curves which leads to Rao's co-variance adjusted estimators, *Biometrika*, **64**, 141–143.

Frish, R. E. and Revelle, R. (1969). The height and weight of adolescent boys and girls at the time of peak velocity of growth in height and weight: longitudinal data, *Human Biology*, **41**, 536–559.

Fuller, W. A. and Battese, G. E. (1973). Transformations for estimation of linear models with nested-error structure, *Journal of the American Statistical Association*, **68**, 626–632.

Ghosh, M., Grizzle, J. E. and Pranab, K. S. (1973). Nonparametric methods in longitudinal studies, *Journal of the American Statistical Association*, **68**, 29–36.

Goldstein, H. (1979). Some models for analysing longitudinal data on educational attainment, *Journal of the Royal Statistical Society*, A; to appear.

Goldstein, H. and Johnston, F. E. (1978). A method for studying shape change in children, *Annals of Human Biology*, **5**, 33–39.

Gould, S. J. (1966). Allometry and size on ontogemy and phylogeny, *Biological Reviews*, **41**, 587–640.

Gower, J. C. (1976). Growth-free canonical variates and generalized inverses, *Bulletin of the Geological Institute of Uppsala*, **7**, 1–10.

Greenhouse, S. W. and Gaisser, S. (1959). On methods in the analysis of profile data, *Psychometrika*, **24**, 95–112.

Grizzle, J. E. and Allen, D. M. (1969). Analysis of growth and dose response curves, *Biometrics*, **25**, 357–361.

Healy, M. J. R. and Goldstein, H. (1976). An approach to the scaling of categorized attributes, *Biometrika*, **63**, 219–229.

Heath, B. H. and Carter, J. E. L. (1967). A modified somatotype method, *American Journal of Physical Anthropology*, **27**, 57–74.

Hills, M. (1968). A note on the analysis of growth curves, *Biometrics*, **24**, 189–196.

Hills, M. (1974). "Statistics for Comparative Studies" Chapman and Hall, London.

Hopkins, J. W. (1966). Some considerations in multivariate allometry, *Biometrics*, **22**, 747–760.

Hudson, D. J. (1966). Fitting segmented curves whose join points have to be estimated, *Journal of the American Statistical Association*, **61**, 1097–1129.

Huxley, J. S. (1924). Constant differential growth ratios and their significance, *Nature*, **114**, 895–896.

Huxley, J. S. (1932). "Problems of Relative Growth" Methuen, London.

Huynh, H. and Feldt, L. S. (1976). Estimation of the Box correction for degrees of freedom from sample data in randomized block and split plot designs, *Journal of Educational Statistics*, **1**, 69–82.

Jenss, R. M. and Bayley, N. (1937). A mathematical method for studying the growth of a child, *Human Biology*, **9**, 556–563.

Jolicoeur, P. (1963). The multivariate generalization of the allometry equation, *Biometrics*, **19**, 497–499.

Joreskog, K. G. (1970). A general method for analysis of covariance structures, *Biometrika*, **57**, 239–251.

Kendall, M. G. and Stuart, A. (1967). "The Advanced Theory of Statistics" Vol. II, Griffin, London.

Khatri, C. G. (1973). Testing some covariance structures under a growth curve model, *Journal of Multivariate Analysis*, **3**, 102–116.

Kidwell, J. F. and Howard, A. (1970). The inheritance of growth and form in the mouse; III, Orthogonal polynomials, *Growth*, **34**, 87–97.

Kleinbaum, D. G. (1973). A generalization of the growth curve model which allows missing data, *Journal of Multivariate Analysis*, **3**, 117–124.

Krause, G. F., Siegel, P. B. and Hurst, D. C. (1967). A probability structure for growth curves, *Biometrics*, **23**, 217–225.

Lee, Y. K. (1974). A note on Rao's reduction of Potthoff and Roy's generalized linear model, *Biometrika*, **61**, 349–351.

Leech, F. B. and Healy, M. J. R. (1959). The analysis of experiments on growth rate, *Biometrics*, **15**, 98–106.

Lozy, M. E. (1978). A critical analysis of the double and triple logistic growth curves, *Annals of Human Biology*, **5**, 389–394.

Marubini, E., Resele, L. F., Tanner, J. M. and Whitehouse, R. H. (1972). The fit of Gompertz and logistic curves to longitudinal data during adolescence on height, sitting height and biocramial diameter in boys and girls of the Harpenden Growth Study, *Human Biology*, **44**, 511–523.

Morrison, D. F. (1972). The analysis of a single sample of repeated measurements. *Biometrics*, **28**, 55–71.

Mosimann, J. E. (1970). Size allometry: size and shape variables with characterizations of the lognormal and generalized gamma distributions, *Journal of the American Statistical Association,* **65**, 930–945.

Nelder, J. A. (1961). The fitting of a generalization of the logistic curve, *Biometrics*, **17**, 89–110.

Nelder, J. A. (1962). An alternative form of a generalized logistic equation. *Biometrics*, **18**, 614–616.

Pearl, R. and Reed, L. J. (1923). On the mathematical theory of population growth, *Metron*, **3**, 6–19.

Pillai, K. C. S. (1967). Upper percentage points of the largest root of a matrix in multivariate analysis, *Biometrika*, **54**, 189–194.

Poirier, D. J. (1973). Piecewise regression using cubic splines, *Journal of the American Statistical Association*, **68**, 515–524.

Preece, M. A. and Baines, M. J. (1978). A new family of mathematical models describing the human growth curve. *Annals of Human Biology*, **5**, 1–24.

Rao, C. R. (1958). Some statistical methods for the comparison of growth curves, *Biometrics*, **14**, 1–17.

Rao, C. R. (1959). Some problems involving linear hypotheses in multivariate analysis, *Biometrika*, **46**, 49–58.

Rao, C. R. (1965). The theory of least squares when the parameters are stochastic and its application to the analysis of growth curves, *Biometrika*, **52**, 447–458.

Reeve, E. C. R. and Huxley, J. S. (1945). Some problems in the study of allometric growth. *In* "Essays on Growth and Form", (Clark, E. E. LeGros, and Medawer, P. B., ed) Oxford University Press, London.

Reyment, R. A. and Banfield, C. F. (1976). Growth-free canonical variates applied to fossil foraminifers, *Bulletin of the Geological Institute of the University of Uppsala,* **7**, 11–21.

Robertson, T. B. (1908). Arch. F Entwicklungsmechanick, **25**, 581–583.

Robertson, T. B. (1915). Studies on the growth of man, I, The prenatal and postnatal growth of infants, *American Journal of Physiology*, **37**, 1–42.

Scammon, R. E. (1927). The first seriatim study of human growth, *American Journal of Physiology Anthropology*, **10**, 329–336.

Scheffe, H. (1959). "The Analysis of Variance", Wiley, New York.

Seigel, S. (1956). "Nonparametric Statistics for the Behavioural Sciences", McGraw-Hill, New York.

Sheldon, W. H., Dupertuis, C. W. and McDermott, E. (1954). "Atlas of Men", Harper Brothers, New York and London.

Sneath, P. H. A. (1967). Trend surface analysis of transformation grids, *Journal of Zoology*, **151**, 65–122.

Tanner, J. M. (1964). "The Physique of the Olympic Athlete", George Allen and Unwin, London.

Tanner, J. M., Whitehouse, R. H., Marubini, E. and Resele, L. F. (1976). The adolescent growth spurt of boys and girls of the Harpenden Growth Study, *Annals of Human Biology*, **3**, 109–126.

Thompson, D'Arcy W. (1917). "On Growth and Form", Cambridge University Press (reprinted 1942).

Weiss, P. and Kavanu, J. L. (1957). A model for growth and growth control in mathematical terms, *Journal of General Physiology*, **41**, 1–47.
Winsor, C. P. (1932). The Gompertz curve as a growth curve, *Proceedings of the National Academy of Sciences*, **18**, 1–8.
Wishart, J. (1938). Growth rate determinations in nutrition studies with the bacon pig, and their analysis. *Biometrika*, **30**, 16–28.
Woolson, R. F., Leeper, J. D. and Clarke, W. R. (1978). Analysis of incomplete data from longitudinal and mixed longitudinal studies, *Journal of the Royal Statistical Society*, A **141**, 242–252.
Wright, S. (1926). Book Review *In Journal of the American Statistical Association*, **21**, 494.
Zerbe, G. O. and Walker, S. H. (1977). A randomization test for comparison of groups of growth curves with different polynomial design matrices, *Biometrics*, **33**, 653–657.
Zucher, L., Hall, L.,Young, M. and Zucher, T. F. (1941). Quantitive formulation of rat growth, *Growth*, **5**, 415–436.

5
The Analysis of Models for Relating Measurements at Different Occasions

We now turn from the models dealt with in Chapter 4, which express a relationship between a measurement and a time scale, to models which directly relate measurements made at one occasion to those made at other occasions. For example, instead of expressing height as a function of age, we can study the relationship between height at, say age 8·0 and height at ages 7·0, 6·0 etc.

There are two main situations where we would want to use such a model in preference to an explicit time-related model. Firstly, the time related models are symmetrical with respect to time. As far as the equations are concerned individuals could just as easily be moving in either time direction. Conditional models, where measurements at some occasions are regarded as dependent on measurements at earlier occasions, do recognize implicitly the direction of time and can therefore be used more directly in the study of causal relationships. In these models the earlier measurements are treated as having fixed values rather than as random variables.

The second distinguishing characteristic of the time related models, as has already been pointed out, is the implicit assumption that the same measuring instrument is used at every occasion. For most kinds of measurements which are not physical ones, however, the same measuring instrument generally cannot be used at every occasion. If we are prepared to assume that different measuring instruments used on separate occasions are, in fact, measuring the same underlying quantity, then we may adopt one of the two techniques discussed in Chapter 3 to derive a common scale of measurement, although little use seems to have been made of the procedure of calibrating different measurements using overlapping age intervals. More use had been made of the separate standardization of measures at different occasions and before describing models relating occasions, we need to devote some attention to the difficulties of this kind of standardization.

To simplify the discussion we shall assume that the measurements have been standardized to have the same normal distributions at each occasion. Suppose we wish to estimate the rates of change between two occasions. We note first of all that the average rate of change for the population is by definition zero, so that we can only study *relative* rates of change for different groups. Another problem arises here, however, because in addition to forcing the distributions at each occasion to have the same mean value, we have also obliged them to have the same form of distribution, in particular the same variance. To see one of the implications of this, we shall consider the case of growth in height between the ages of four and eight years. Using data from British growth standards (Tanner *et al.*, 1966) we see that the mean difference in height between boys and girls remains constant at 1·2 cm. The total population standard deviation of height, however, increases by a third, from 4·34 cm to 5·77 cm. Thus if we were to standardize height to give the same population standard deviation at each age we would find that the mean difference at age eight was only three-quarters the mean difference at age four.

While it seems natural, in the case of height, to work in terms of the same absolute unit at each age, the ultimate justification for choice of unit has to be based either on considerations of theory or of practical utility. For example, if we find that using standardized height gave simpler descriptions of relationships with other factors and predictions, then we might wish to use such a transformed measurement instead of the original one, although we should also have to weigh up the disadvantages of using such an unfamiliar scale. In fact, for technical reasons, statisticians often seek variance-stabilizing transformations which are designed to ensure that all measurements, for all values of other factors, have the same standard deviation.

Similar considerations apply to measurements such as reading attainment, with the added problem that we generally have no absolute scale to start with. Whereas with height the growth rate has a well-understood meaning, it is not at all clear that the use of standardized reading scores resolves the problem of attaching a meaning to these measurements. We see then that for measurements of this kind we are obliged to look for other approaches to the measurement of change, and the remainder of this chapter is concerned with models for this. We shall assume that the variables used in these models have clear interpretations along the lines described in Chapter 3; namely that they are the variables *appropriate* to each occasion without necessarily being measures of the *same* underlying attribute.

We begin our discussion by considering two occasions with a single measurement at each one.

5.1 Two-occasion Models

Consider Fig. 5.1 which is a scatterplot of reading test scores made at one occasion against reading test scores made at a second occasion. For the moment we consider the case where the same instrument is used at each occasion.

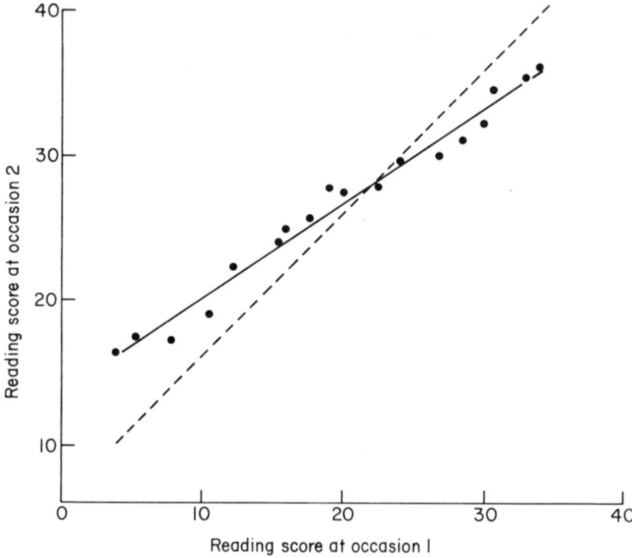

Fig. 5.1 Hypothetical scatterplot of reading score at occasion 2 by reading score at occasion 1. Unbroken line is fitted regression line of occasion 2 score on occasion 1 score. Broken line is line with unit slope.

The relationship between the two test scores in this case seems to be adequately summarized by a straight line, as drawn. In many situations, of course, more complex relationships will exist, necessitating either the fitting of more complex curves, such as polynomials, or the making of suitable transformations of the scales to produce linear relationships between the transformed measurements. We have already seen, when studying simple changes in the same or similar measurements between occasions, that the choice of transformation may present quite a difficult problem. We shall discuss how different types of transformations may change both the form of relationships between measurements and the resulting inferences.

Before considering this, however, we note an important point arising from this diagram. If the equation of the continuous line is written as

$$y_i = \alpha + \beta x_i + \varepsilon_i \tag{5.1}$$

then the broken line has an equation of the form

$$y_i = \alpha + x_i + \varepsilon_i \tag{5.2}$$

i.e. where $\beta = 1$.
In this case (5.2) appears to reduce to the analysis of simple change scores, that is, a time related model which could be written as

$$y_i - x_i = \alpha + \delta_i. \tag{5.3}$$

The x_i in (5.2), however, are not random variables as in (5.3) but are considered as having fixed values. The analysis is still a conditional one, therefore, and the change of scale does not affect this. Indeed, none of the important conclusions from an analysis of the model is altered by a change of scale, unlike the simple change score model, as illustrated by the height differences example.

It should also be pointed out that the situation illustrated in Fig. 5.1 is unlike the bivariate allometry case considered previously, where both variables had the same status. Here the values of the second occasion variable are studied for *given* first occasion variable values so that the situation is essentially an asymmetric one. We shall discuss this point further in Section 5.1.4.

We should always begin with a study of the actual form of relationship present in the data. Of course we may be guided by theory or past evidence on this, and we will want, where feasible, to obtain as simple a summary as possible. In certain situations there may indeed be guidance on this count, but the only *general* approach is to seek that summary, or model, which best describes the observed data. The problem of finding the appropriate model for the prescribed data is a common one in applied statistics, and this point needs to be emphasized because there has sometimes been a certain amount of discussion (see e.g. Werts and Linn, 1970) as to whether an equation of the form (5.1) (adjusted change) or (5.3) (simple change) is appropriate to use. In essence this is an irrelevant argument since (5.1) makes different assumptions from (5.3) which *may* apply in some situations, but whether or not it does apply is a matter of empirical fact. It is indeed a relatively simple matter to demonstrate how, with a given set of data, the use of (5.1) or (5.3) can lead to very different conclusions when, for example, different groups are being compared (Lord, 1967). For a general discussion of the problem of model fitting, the reader is referred to Blalock (1971).

5.1.1 Regression to the Mean

With the usual assumptions made in simple linear regression, namely that a straight line is appropriate, and that the deviations about the line belong to a single normal distribution, then in the usual way we can estimate the parameters of the line and assign confidence intervals to them. We can also use these estimates to predict the second occasion measurement from the first occasion measurement. Points lying on the regression line represent the average value of second occasion measurements for given first occasion measurements. If we examine Fig. 5.1 we notice the following: the mean value of reading score at the first occasion is 22·0 and at the second occasion is 28·0, an average increase of six score points. For reading scores at the first occasion which are below the mean, the average second occasion score is more than six points higher. For a second occasion score of 10·0, for example, the average second occasion score is 20·0, an increase of 10·0 points. For scores at the first occasion which are above the mean, the increase is less than six points. These results are simply a reflection of the fact that the slope of the fitted line is less than 1·0. If it were greater than 1·0 then the reverse would hold.

The previous paragraph has been concerned with single scores and average values at each occasion, but in order to begin to assess the importance of differences between these we need to be able to relate them to the overall variation between individuals. Suppose, for simplicity, that the reading score distributions at each age are normal so that the variation of scores can be summarized by the standard deviation. Suppose also that the standard deviations at each occasion are equal. From the discussion in the previous paragraph it is easy to see that for all those scores at the first occasion which are below the first occasion mean, their expected value at the second occasion is closer (in terms of the standard deviation) to the second occasion mean. Thus a score of 10·0 at the first occasion is 12 points below the mean, whereas its expected value at the second occasion of 20·0 is only eight points below the mean, i.e. in terms of the standard deviation it is two-thirds the distance from the mean. The reverse holds for first occasion values above the mean. This situation is known as "regression towards the mean" and is, in fact, found with almost all measurements. Galton (1886) pointed out this phenomenon with respect to the relationship between the height of adults and the height of their parents. He showed that the offspring of relatively tall parents had heights, on average, closer to the population mean than the heights of their parents (actually the average of the maternal and paternal heights), the value of β being about two-thirds. It so happens that the height distribution for Galton's adults and their parents had approximately the same standard deviations. Had the standard deviation of the offspring been markedly greater than that of their parents, this phenomena might not have been observed. It is easy to see how this can happen if we stretch the scale of the second occasion measurement in Fig. 5.1 by multiplying every value by 2. Using the new scale the value of β would be doubled and would now be greater than 1·0. Also the average second occasion measurement would now be *further* from the mean for a low first occasion measurement. Relative to the second occasion standard deviation, however, it would be just as far from the mean as before. Thus the significance of any regression to the mean will lie partly in the scales of measurement used. If the distributions at each occasion are manipulated (as is done with many measurements) to have the same standard deviation at each age, then, in fact, regression will almost certainly occur.

This can be understood more simply as follows. When the standard deviations at each age are the same, the regression coefficient β is equal to ρ, the correlation coefficient between the measurements. Since ρ is always less than or equal to 1·0 (we need only consider positive values), it follows that $\beta \leqslant 1{\cdot}0$ and regression to the mean will always take place, except in the exceptional circumstance that $\rho = 1{\cdot}0$.

We see, therefore, that an explanation of this phenomenon may be sought in an explanation of the reasons for a less than perfect correlation. We shall deal later with one of these reasons, namely measuring error and see how its influence can be allowed for. A more extended discussion of regression to the mean is given by Healy and Goldstein (1978).

5.1.2 Differences Between Groups

Before discussing models for the analysis of data from different groups, we shall describe briefly the ways in which such data can arise since these will influence the interpretations which can be made.

Imagine several groups of individuals all measured at one occasion. These might be children from different social classes, or for example children to whom different treatments such as teaching methods are subsequently applied. Two types of situations can be distinguished. The first is where groups remain intact at each occasion and interest centres on estimating change between occasions. This is analogous to the example in Chapter 4 where growth curves were fitted to heights of children grouped by the heights of their parents. The second situation is where a second occasion measurement is meant to measure response to a treatment and the first occasion measurement is made prior to the treatment in order to adjust for bias and/or to increase precision. This second case was discussed in Chapter 1 where a distinction was drawn between randomized assignment at the first occasion and non-random assignments. A further distinction in the latter case was between quasi-experimental and observational studies.

Fig. 5.2 illustrates a possible situation with three randomly assigned reading schemes. The equations can be written

$$y_{ij} = \alpha_j + \beta x_{ij} + \varepsilon_{ij} \qquad j = 1,\dots k \tag{5.4}$$

where k is the number of treatments and i refers to individuals within treatments. The formal analysis of (5.4) known as the analysis of covariance, can be found in any standard reference, for example Scheffe (1959).

With a non-randomized experiment the first occasion expected mean values are not equal in general and the interpretation of the diagram will change.

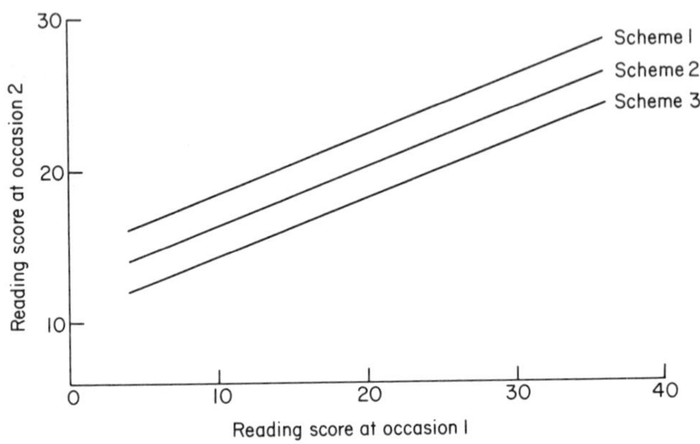

Fig. 5.2 Outline of hypothetical randomized experiment to compare three reading schemes. Parallel regression lines have been fitted to summarise the relationships.

As well as interpreting equations (5.4) as models for the situation where different treatments are applied to different groups, these equations could represent the relationships between measurements made at two time points without in any way attempting to alter or treat the groups. Such a situation is nearer to the ordinary observational study than is either the randomized experiment or the quasi-experiment discussed in Chapter 1. Hence its interpretation will tend to differ and the fact that we may use the same statistical model in each case should not make us forget to distinguish carefully between the different motivations behind these models. In the following example we shall examine the relationships between reading scores at two occasions for three groups of children, and when we discuss the interpretations of the results it will be possible to see where alternative interpretations might apply in the randomized or quasi-experimental case.

Reading scores and the social class of each child's father are available on over 9000 children in the British National Child Development Study (N.C.D.S.) at the ages of seven and 11 years (Fogelman and Goldstein, 1976). The original analysis studied additional factors but we restrict ourselves here to comparing the reading score relationships for three groups of children defined in terms of their social class group at the age of seven; those whose fathers were in non-manual occupations (Social Class Non-manual), those with fathers in skilled or semi-skilled manual occupations (Social Class III Manual and IV) and those with fathers in unskilled manual occupations (Social Class V).

In the original analysis the seven and 11 year scales were graduated in "years of age", one year representing the average change in score over a year of age, at seven and 11 years respectively. In practice, however, the relationship between the test score and age is non-linear, so that such a conversion is approximate, especially at the extremes of the scales. An alternative to the use of the original arbitrary scale units is to transform the scores at each age so as to have as near as possible the same distribution. The 11-year reading score has been transformed to have a normal distribution with zero mean and unit standard deviation, and the seven-year scale has been transformed to provide an approximately linear relationship with 11-year score and then scaled to have a zero mean and unit standard deviation. Further details of these transformations are given in Goldstein (1979) and the following data are taken from that paper.

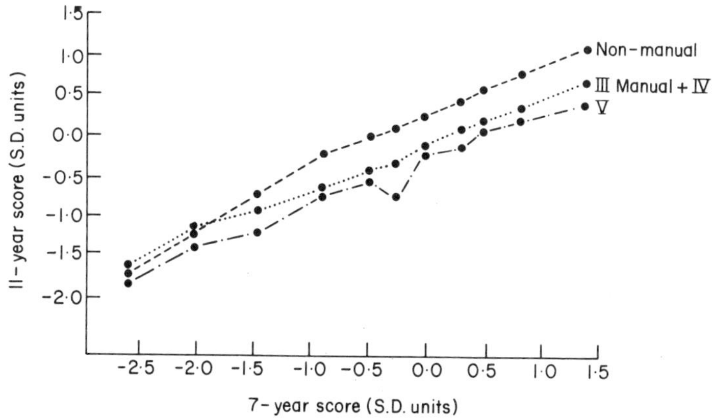

Fig. 5.3 Mean 11-year reading scores for three seven-year social class groups for seven-year reading scores. (NCDS data.)

Fig. 5.3 shows the mean 11-year score for each social class for the range of seven-year scores. The general picture is clear. For each seven-year score, children in Social Class V have a lower mean 11-year score than other children. Except for the lowest seven-year score, children in Social Class III manual and IV have a lower mean 11-year score than non-manual children. Hence, ignoring for the time being the problem of measurement error or reliability (see Section 5.2), we may form the tentative conclusion that children exhibiting the same attainment at seven, on average do better at 11 the higher their seven-year social class. Thus, we may conclude that the change in reading attainment between these two ages differs between these social class groups.

In spite of this general conclusion there do appear to be some irregularities in the pattern. For low seven-year scores there is a "cross over" of the upper two lines. The number of cases involved, however, is small and the estimates of the means have large sampling errors. To avoid this problem associated with small numbers of cases we can summarize the relationship by fitting straight lines through the mean 11-year scores. Figure 5.4 shows these linear regression lines for each group.

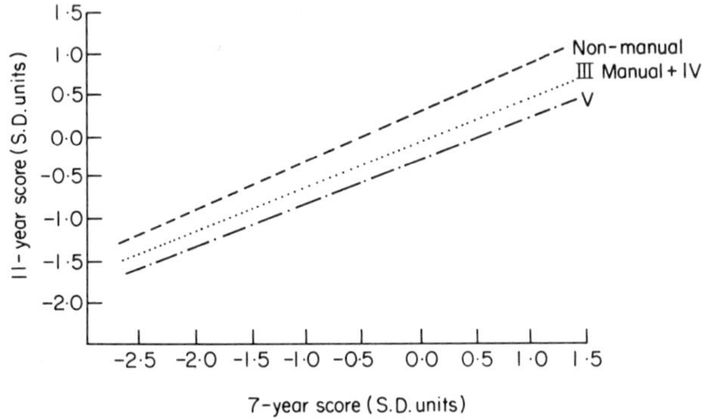

Fig. 5.4 Fitted regression lines of 11-year reading score on seven-year reading score for three seven-year social class groups (NCDS data).

We have now eliminated the cross over, and although there is still a slight interaction or increasing difference in mean 11-year scores with increase in seven-year score, it is not statistically significant. We note that for the present purpose the point of the low-order curve fitting is to make simple general statements about the relationships. If we wished accurately to *predict* 11-year score given seven-year score then the precise form of curve would become more important.

Where there are large differences between the slopes of such regression lines we might still for simplicity constrain the lines to be parallel and carry out a standard analysis of covariance. We could, however, average the group differences in other ways. We might for example, ignore extreme scores if these were causing the non-parallelism or alternatively we could form a weighted average of the group differences using the distribution of the seven-year scores for just one group, say V, as a standard population. This latter possibility may be appropriate in a randomized or quasi-experimental situation with, say, two groups where one of the groups is a control group and one a treatment group. If the slopes of the regression lines are unequal, then it may be reasonable to estimate the average difference over the range of values present in the treated group, i.e. weighting with respect to the distributions of x-values in the treated group. In this case, it turns out that the slope used to calculate the average difference is that estimated from the control group.

Cochran and Rubin (1973) discuss this and other cases in some detail. In practice, as in the present example, we can often avoid such complications by finding a suitable transformation to eliminate such interaction effects.

In our example, when lines of equal slope are fitted, the mean difference between 11-year scores for given seven-year scores is 0·37 units between Non-manual, and III Manual + IV groups and 0·19 units between III Manual + IV and V groups. The pre-existing mean differences at the age of seven are respectively 0·50 and 0·43 units, so that the additional differences are somewhat smaller than the pre-existing differences, both pairs of differences being measured in terms of their respective population standard deviations.

A further problem of interpretation arises in this example because of the different reading tests used at the two ages. At the age of seven the test essentially measured the ability of the child to recognize words, whereas the 11-year test measured comprehension. It may seem difficult to talk of a change in reading attainment if different aspects of reading are being measured. As we saw in Chapter 3, this difficulty is not necessarily resolved by using the same test at each age, since the test which is appropriate at one age may not be appropriate at another age.

The question of appropriateness is crucial. We might, for example, decide that a test is appropriate if at each age it measured an ability which educationalists considered appropriate and which, for example, formed the basis of school placement. Our test would then have been designed to measure aspects of the effects of the educational system. We might alternatively be able to decide appropriateness on the basis of theoretical arguments concerning educational development. In this case we would be able to give a meaning to statements of the kind; "for children who had reached the same given level of attainment in terms of development A, those in group 1 had reached a higher level of development B at a subsequent age." Whichever of these approaches we adopt, none of them would be applicable to the analysis of simple differences between (standardized) scores, such differences having no ready interpretation.

For randomized or quasi-experimental studies, measurements should be specifically related to the aims of the study. Often, especially where the duration of the study is short, the same measurement might be used at each occasion where the purpose of the study is to assess the change in a particular measurement. Even here, however, care is needed to ensure that the effects of natural maturation are properly accounted for in deciding on the appropriateness of the measure.

5.1.3 Matching

If we wish to estimate the difference between two groups at a second occasion, we can match individuals on the first occasion measurement. That is, we choose pairs of cases with the same first occasion measurement (x) and study the differences for the second occasion measurement (y). If this can be done, it is clear that we can draw the same kinds of inferences as above. We can study the dependence of the differences on x, e.g. whether there are non-parallel regression slopes. We may have similar problems to those of the previous section, for example in deciding how to average the differences if the original population from which the groups are sampled have different x distributions. There are also practical problems in deciding how closely in agreement the x values of a pair have to be in order to be counted as a match. There are also problems with this approach when errors of measurement in x are present. Matching may be particularly useful in quasi-experimental studies when it is carried out for other confounding values as well but in general it seems to have no advantage over the model fitting approach although a combination of matching and regression line fitting can sometimes be useful. A detailed discussion and comparison of methods is given by Cochran and Rubin (1973).

5.1.4 Time Reversed and Functional Relationship Models

In relating two occasions we have so far studied only the relations between the mean value of the second occasion measure (y) and the first occasion measure (x). This arose naturally from the formulation of the problem which specifically acknowledged the time direction from occasion one to occasion two. We could, however, also view the problem in two other ways. For example, we could ask about mean differences in first occasion measures for given second occasion measures. Thus in the above example we would want to know whether, for those children with the same 11-year reading attainment, those in the non-manual social class had higher or lower mean seven-year scores than those in, say, social class V. We could interpret this as asking whether, in order to achieve a given 11-year score, children from the non-manual social class have exhibited a greater change than those in social class V. We are attempting here to answer a different question from the previous one, but one, which nevertheless, may be of interest in some situations.

A third approach to studying the relationship between measurements at two occasions is to assume symmetry between the occasions. In equation (5.1) the first occasion measure, x, was treated as having a set of fixed values and the second occasion measurement, y, was defined in terms of its average values for given x values together with a residual term ε which was regarded as a random variable. For the relationship of x on y the reverse holds, y being treated as having a set of fixed values with the mean values of x being studied. We can, however, formulate the relationship so that x and y are treated alike, as follows:

write

$$x_i = X_i + d_i$$
$$y_i = Y_i + e_i \tag{5.5}$$

and

$$Y_i = \alpha + \beta X_i$$

Here, d_i and e_i are random variables, whereas Y_i and X_i are fixed although unobservable. Thus we are postulating an average underlying linear relationship about which individuals vary at random. The variables d_i, e_i incorporate individual variation about the line as well as measurement error if this is present. This functional relationship between x and y was referred to earlier when we discussed allometry where it was interpreted as representing a common growth process underlying two measurements. In the present context we can choose to interpret it as the relationship underlying two measurements at different occasions and perhaps go on to estimate the relationships for different groups of individuals. One of the problems with this approach, however, is the need to define a common scale of measurement. As with growth curves, the time symmetry forces us to impose what is the equivalent of a common scale. This becomes apparent when we attempt to estimate the parameters of this model (α, β and the variances and covariances of d_i, e_i). When the residuals have normal distributions, the usual method (see e.g. Kendall and Stuart, 1967) involves specifying a predetermined value for the ratio of the variances of the d_i and e_i. Thus the estimates obtained will depend on the assumption we make about the relative variability of the two scales of measurement.

5.1.5 Cross-lagged Panel Correlations

We have emphasized the importance of careful model specification for the analysis of longitudinal data. We shall now discuss an approach which has been advocated (e.g. Campbell, 1963) as a useful summary method for detecting the direction of influence or causation for two-occasion studies. The point of this discussion is to show how important it is to specify a precise model, and how it is possible to derive erroneous conclusions through the absence of such a model.

Consider measurements of reading, x, and verbal ability, y, attainments, each made on two occasions on a group of children. Suppose we are interested in whether level of performance in verbal ability is determined by level of performance in reading or vice versa. Phrased in this way the question is somewhat unclear, but it has been suggested that directions of causation can be inferred by comparing "cross-lagged" correlations between occasions. We write $r(x_1, y_2)$ as the cross-lagged correlation between the x measurement on occasion 1 and the y measurement on occasion 2, and correspondingly for $r(x_2, y_1)$. The principal direction of causation is said then to go from x to y if $r(x_1, y_2) > r(x_2, y_1)$. The justification for this reasoning seems to lie in the apparently reasonable statement that if x on occasion 1 is more strongly associated with y on occasion 2 than vice versa, then x is determining y rather than the other way around. Concentration on these correlations, however, ignores any other relationships between x and y, and the following simple model illustrates how this can be misleading.

Let us suppose that reading attainment is determined by verbal ability rather than the other way around. We specify precisely what this means in the following equations (the residual terms are omitted for convenience).

$$\begin{aligned} x_1 &= \alpha_1 + \beta_1 y_1 \\ x_2 &= \alpha_2 + \beta_2 x_1 + \gamma y_1 \\ y_2 &= \alpha_3 + \beta_3 y_1 \end{aligned} \tag{5.6}$$

where x, y are standardized to have variances equal to 1. We see that whereas x_2 is related to both x_1 and y_1, y_2 depends only on y_1.

We have after a little manipulation, the correlations

$$r(x_1, y_2) = \beta_1 \beta_3$$

$$r(x_2, y_1) = \beta_1 \beta_2 + \gamma$$

and

$$r(x_1, y_2) - r(x_2, y_1) = \beta_1(\beta_3 - \beta_2) - \gamma.$$

We can now give the parameters of these equations arbitrary values, consistent with the correlations lying between 0 and 1.
For example:

$$\beta_1 = 0{\cdot}5, \beta_3 - \beta_2 = 0{\cdot}25, \gamma = 0{\cdot}1 \text{ giving } r(x_1, y_2) - r(x_2, y_1) = 0{\cdot}025$$

which is greater than zero and according to the cross-lagged correlation argument would imply a direction of causation from x to y, whereas equations (5.6) show the reverse to be the case. In fact, it is *only* by writing down an explicit model that we can give any real meaning to statements about directions of causation, remembering, of course, that in dealing with observational data the term "causation" should be used cautiously. Kenny (1973) describes a simple factor analysis model to demonstrate a similar argument.

While on the subject of model specification, a few general remarks are appropriate. In many situations, several different plausible models are possible, each leading to different conclusions and each also being a reasonable fit to the available data. This book is not the place for a detailed discussion of how to proceed in these circumstances and the reader is referred to other texts such as that of Tukey (1976). We shall see in the following sections how the number of possible models can increase when measurement error is introduced and when more than two occasions are considered. Some of these models can become quite complex, often to the point where it is unlikely either that appropriate data or substantive theory exist to distinguish between alternatives. In such a situation, simplicity is often a good guide. In discussing these further topics, we limit ourselves to presenting simple general approaches, making references where appropriate to other discussions of the more complex models.

5.2 Measurement Errors

Most measurements on individuals are inexact in the sense that repeated application of the measuring instrument, even under standard conditions, will yield somewhat different values. Such an instrument is said to be fallible. This measurement error is present in both physical and behavioural measurements, although typically smaller in the former as well as being easier to quantify. If we repeatedly measure a child's weight in a short space of time in standard conditions, for example, then we will obtain a series of different measurements whose values will fluctuate in some apparently random manner. Slight uncontrollable differences in measuring conditions, and short term fluctuations in the weight itself will both contribute to this variation. If we imagine taking many such measurements, we would be able to describe their distribution, and in particular this distribution would have an average value which we can regard as estimating the individual's "true value". For simplicity, we assume a symmetric distribution and write

$$x_i = X_i + e_i$$

where X_i is defined as the true value, e_i is the measurement error and x_i is the measured value. In practice our estimate of measurement error will depend on what is defined to be random error and what can be classified systematically as a component of the basic measurement itself. For example, it would be possible to relate a child's weight to the time of day when it was measured and find, say, a cyclical variation. Measurement error would then consist of whatever individual variation was left between measurements after allowing for the effect of time of day. Alternatively, we might decide to include the "between time of day" variation in the total measurement error. The choice depends on what questions we are seeking to answer and what practical options are available. A more detailed discussion of this point in a particular context can be found in Fletcher *et al.* (1976).

There are some situations where the actual measured values are of interest and we wish to work entirely with these. In other situations, however, we wish to make inferences about individuals' true values and since, as we shall see, we may obtain different inferences in the two cases, it is important to study the effect of allowing for measurement error. To illustrate the effect measurement error can have, we describe a simple situation where a group of individuals is measured on two occasions, close enough together so that the true values remain unchanged, and any differences are due solely to measurement errors. One word of caution is necessary. In practice, especially with mental tests, it is not possible *independently* to measure an individual repeatedly at short intervals, since the result of one measurement influences or is carried over to the value of subsequent ones. Thus, an estimate of measurement error usually has to be obtained indirectly. In the following discussion we shall assume that the errors are, in fact, independent. Thus the situation we describe may be somewhat artificial but, nevertheless, is useful in illustrating a general point. Furthermore, if we extend the time interval, measurement errors will become more independent and the true values themselves will change. The following discussion gives a base line so that a model dealing with changes in true values can be superimposed on the results obtained.

Consider a measurement made at two closely spaced occasions. If the errors are independent with the same distribution at each occasion, with the same true values then the correlation between the measurements is

$$R = \frac{\sigma_X^2}{\sigma_X^2 + \sigma_e^2} \tag{5.7}$$

where σ_X^2 is the between individual variance of the true values and σ_e^2 is the variance of the measurement errors. The parameter R is known as the reliability of the measurement. A discussion of the theory of measurement errors in mental measurement together with methods of estimating R etc. can be found in Lord and Novick (1968).

If we denote the measurement at occasion 2 by y and at occasion 1 by x, and if we assume for simplicity that the means are both zero, then we can write the regression of y on x as

$$y = Rx \tag{5.8}$$

where, in general, $R < 1$.

It follows from (5.8) that if $x > 0$ then $y < x$ and if $x < 0$ then $y > x$. In other words, the expected or average value of the second occasion measurement for a given 1st occasion measurement is closer to the mean (zero) of the whole distribution. Likewise, if we consider a group of individuals defined as having values greater than a given value on the first occasion, their average value on the second occasion will be lower than their average on the first occasion. The phenomenon is mathematically equivalent to the "regression to the mean" discussed earlier, but the "regression effect" observed here results solely from the characteristics of the measuring instrument rather than from the relationship between individuals' true values on each occasion. A corollary of this is that when we observe a regression to the mean on a group of individuals measured on two occasions, we cannot, without further information, determine how much occurs as a result of the relationship between true values and how much results from measurement error. A more detailed discussion of this with examples is given by James (1973), who presents formulae for estimating and separating out the two effects. The problem is there discussed in a clinical context where subjects for treatment are selected on the basis of having measurements greater than a certain value and are then remeasured after treatment. In the absence of a control group not undergoing treatment, it is necessary to decide whether any regression to the mean is a result of treatment or simply measurement error. By making assumptions about the distributions and about some of the parameters, James shows how the required estimates can be obtained. In particular he shows that if the reliability of the measurement is low, then a major part of the shift towards the population mean can be due to measurement error, when, in the absence of measurement error, a treatment applied before the second measurement actually causes a large true proportional shift toward the mean. Table 5.1 illustrates this, where y and x are standardized as before. The increasing importance of measurement error, as the reliability decreases, is clear.

We now turn to the effect of measurement error on the group comparison models discussed in Section 5.1.2. We assume that interest centres on relationships between true values rather than between the measurements themselves. This is reasonable in many applications where we would not wish the nature of our inference to change if we started to use a more reliable instrument. There is one kind of situation, however, when it would be appropriate to use observed values, namely when we are interested in predicting a second occasion value, the former always being measured by

Table 5.1 Contribution of Measurement Error to Regression to the Mean with a true proportional Change between Occasions of 80%.

Reliability (R)	% regression to mean due to measurement error
0·4	65
0·5	56
0·6	45
0·8	24
0·9	12

the same fallible instrument. In this case, the best prediction equation is that using the actual measurements made. We might also want to use measures with given reliability values, where decisions or judgements about individuals were based on such measures; for example, the allocation of a school or university place. The true-value model can then serve as a baseline. We need to be careful about the use of the phrase "true value", especially with mental measurements. As we have defined it, it refers to an average value associated with a particular measuring instrument. Thus, for example, an individual's "true" reading attainment may depend on which reading attainment test is being used. Other writers have suggested, however, that any given test uses a sample of items from a much larger possible set and that an individual's true value should be measured with reference to this larger set. Gleser *et al.* (1965) discuss such issues.

Consider again the population simple linear regression relationship

$$y = \alpha + \beta x$$

where x is measured with some error. (Measurement error in the dependent variable y has no biasing effect on the parameter estimates of interest, serving only to increase the residual variance about the regression line.) We are interested in the relationship between the true values X and Y.

$$Y = \alpha' + \beta' X$$

If β' is positive then, when measurement error is present, $\beta < \beta'$ (Kendall and Stuart, 1967). The absolute value of the slope of the regression line is always underestimated when using observed values, and the estimate of the slope is called "inconsistent". This means that, irrespective of sampling variation, even when the sample becomes very large its expected value is always an underestimate of the true value. The relationship between the population values β and β' is

$$\beta' = \beta/R \tag{5.9}$$

where R is the reliability of the x measurement. Hence an obvious method of correcting the sample regression slope is to divide it by the reliability. The "correction for attenuation" can be derived by considering the usual estimate for the slope of the regression line, which is

$$\frac{\text{covariance }(y, x)}{\text{variance }(x)}.$$

We require the estimate of true slope

$$\frac{\text{covariance }(Y, X)}{\text{variance }(X)}.$$

The covariance terms are equal and we have (from 5.8),

$$\text{variance }(X) = R \text{ variance }(x).$$

Hence by substitution the required estimate is as above.

Since the covariance of x with any other variable is the same as the covariance of X with that variable, this suggests a general procedure for correcting regression estimates where there is more than one independent variable. In any multiple regression or analysis of covariance we simply use the above formula to obtain the estimated variances of the true values, and use these in the calculations instead of the variances of the observed values. This then gives consistent estimates for the population parameters. A detailed discussion of this procedure is given by Warren *et al.* (1974) and Fuller and Hidiroglou (1978). It depends on having an estimate of σ_e^2 or R which may not always be available, at least not very precisely. We shall discuss this problem further when we consider an example.

Unfortunately, consistency is a rather weak statistical property, only guaranteeing a good estimate with a large sample. Richardson and Wu (1970), however, show that the bias in the regression slope using (5.9) is approximately $2(1 - R)/n$, where n is the sample size. Hence for values of R greater than about 0·5 a sample size of about 100 would guarantee a bias of less than 1%, although as the number of independent variables in the model increases a larger sample size will be necessary.

To illustrate how inferences from an analysis can substantially be altered when measurement error is taken into account, consider Fig. 5.5.

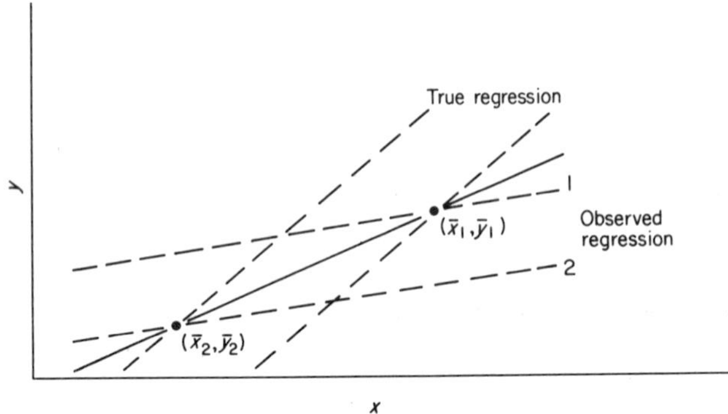

Fig. 5.5 Illustration of inference reversal after a correction for attenuation.

The second occasion measurement (y) has a linear regression on the first occasion measurement (x), in each of two groups, with parallel slopes. The lines through the observed values are the broken lines labelled "observed regression". Each, of course, goes through the mean of each group as shown. We see that, for a given x value, individuals in group 1 have a higher average y value than those in group 2. If we apply the same reliability correction for each line so that the slope is increased to a value greater than that of the line joining the group means we obtain the broken lines labelled "true regression" which still pass through the group means. These lines are parallel, but now, for a given x value, those individuals in group 2 have a higher mean y value than those in group 1, a reversal of the previous inference. Now we go back to the previous example and see what happens when we allow for measurement error in the seven-year reading score.

The reliability of the seven-year score is not known with any precision, but has been estimated to be 0·79 (Goldstein, 1979). If we carry out the adjustment procedure described in the previous paragraph we obtain a new set of parameter estimates. The slope of the regression line (equal for each social class) is increased from 0·61 to 0·79. The difference between the Non-manual Social Class children and those from Social Class III Manual + IV groups is decreased from 0·37 units to 0·27, and that between children from Social Class III Manual + IV and V groups from 0·19 to 0·11 years, for given seven-year score. Thus the social class differences are reduced by about 25% and 40%. It is informative to see how changes in the reliability affect the estimates. Figure 5.6 shows how the difference between the Non-manual and Social Class V groups changes as the reliability changes.

When the reliability reaches about 0·5 the social class difference becomes zero, and below this, social class V has a higher mean than Non-manual for a given seven-year score. If we accept that the true value of the reliability lies

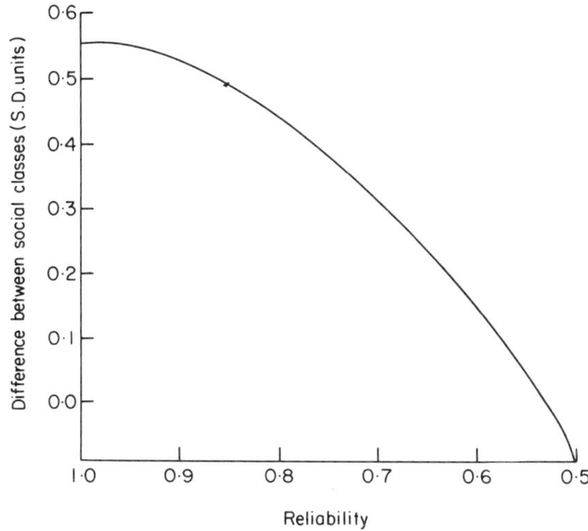

Fig. 5.6 Difference between constants fitted for Non-manual group and Social Class V group in analysis of 11 year on seven year reading score, after adjusting for different reliability values (NCDS data).

in the range 0·75 to 0·85, the social class differences do not change very much in this range. Also, in this example, the social class differences remain statistically highly significant. The direction in which parameter values change will depend on the relative position of group means and the correlations between independent variables as well as on the value of the reliability, and no simple general rules can be given. Often, however, inspection of a diagram such as 5·5 together with simple adjustments to the slopes will give an idea of the approximate effect of a reliability correction. Because the reliability value is estimated from a sample, it will have some sampling variation associated with it. This uncertainty will lead to an increase in the standard errors of the estimated coefficients which can be large if the sample used to estimate the reliability coefficient is small. For example, in the above example the standard error of the estimate of the slope of the regression line of 11 on seven-year reading score is increased by 30% after adjusting for reliability. Nearly three-quarters of this, however, is due to the sampling variability in the reliability estimate. While the total sample size is about 9000 the reliability is estimated using the equivalent of only about 3000 and thus has a relatively large effect on the standard error (Goldstein, 1979).

One method of estimating the reliability of an instrument is to use it twice within a short time-interval on the same individuals. As we noted earlier, this is often impractical because of carry-over effects which mean that the measurement errors are correlated. A common alternative method with mental and behavioural tests is to split the set of items comprising the test into two, and use a formula based on the correlation between the "split halves" to estimate the reliability (Lord and Novick, 1968). Sometimes it is possible to find a second measuring instrument which is highly correlated with the first, measuring a similar although not necessarily identical attribute. If the measurement error of this instrument is independent of that of the first instrument, it is possible to utilize measurements made with the second instrument to obtain corrected parameter estimates, or equivalently an estimate of the reliability of the first instrument. The reason that ordinary regression analysis gives an inconsistent estimate of the true slope lies in the fact that the observed value x is not independent of the measurement error e (the correlation between them $= 1 - R$). If we can replace x by an estimate of x which is independent of e then we will obtain a consistent estimate of the slope and the other parameters. If we choose a second measurement, say z, which is correlated with x we obtain a "least squares" prediction equation for x, for example

$$x_i = a + bz_i \tag{5.10}$$

We may now use this equation to give a predicted x_i value for each z_i and substitute these predicted values in the original equation relating y and x. This procedure is known as 2-stage least squares. If z_i, in fact, comes from a second application of the first measuring instrument, then this gives the usual corrected estimate of slope (5.9) using the observed z_i as the independent variable, with the estimate of R coming from the regression of x on z.

The variable z is called an "instrumental" variable and its use is solely to enable us to obtain consistent estimates. Clearly, having obtained a consistent estimate we can obtain an estimate of the reliability by comparing it with the uncorrected estimate of the regression slope. A detailed discussion of this technique can be found in Johnston (1972). It is by this means that the value of 0·79 for seven-year reading was found and, as mentioned, it has an accuracy equivalent to that found using a more conventional method with a sample of size about 3000. For its success, the instrumental variable method relies on the assumption that the distribution of z is independent of e, the measurement error. If we wish to have an instrumental variable possessing a high correlation with x, however, it may not be possible to guarantee that it is independent of e. For example, with reading attainment we might use a test of general ability or a teachers' rating of reading ability or a combination of both, but if these measurements are obtained at the same time as the reading test itself, they may possess measurement errors which possibly are correlated with those of the reading test.

We see, therefore, that large measurement errors can create serious problems in an analysis, and this is particularly so when no very precise estimates for the value of the reliability are available. It is advisable, before attempting to use a measuring instrument, to try to ensure that its reliability is as high as possible, and that in any case a good estimate of the reliability coefficient is available, based on as large a sample as possible. It will often be found that reliability values quoted for tests are, in fact, based on relatively small samples. Since the size of the standard errors (and the sensitivity of significance tests) in the analysis may depend markedly on the contribution of such reliability values, it may be desirable to make fresh estimates of them.

For technical reasons, a distinction needs to be drawn between the use of reliability values and the use of estimates of σ_e^2. This is discussed in Goldstein (1979).

5.3 Models for More than Two Occasions

In moving from two to three occasions, we at once increase the possible complexity of the models used to represent relationships. As an example we will extend the reading test data by incorporating a further measurement at the age of 16 years.

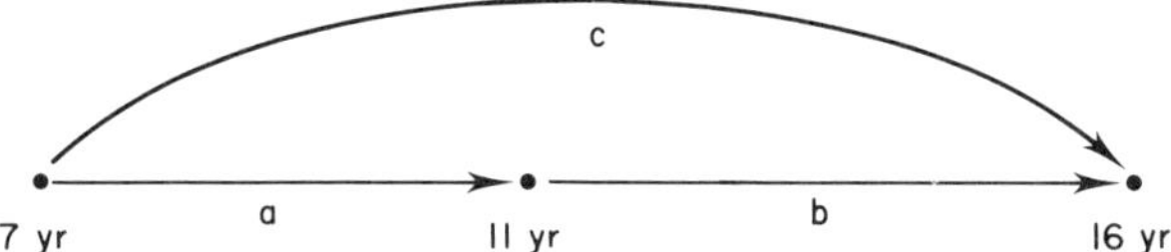

Fig. 5.7 Path diagram for relationships between reading test scores at seven, 11 and 16 years.

We have three possible basic relationships to consider; that from seven to 11 years as before; that from 11 to 16 years, and that from seven to 16 years. It is useful to represent this diagrammatically, Fig. 5.7. The arrows represent these relationships or pathways, and are labelled *a*, *b*, *c*, each pointing from the independent to the dependent variable. For each dependent variable we can write down a model for the relationship, which, for simplicity, we shall assume is linear. For the 11-year measurement we write

$$y_i = \alpha_1 + \beta_1 x_i + e_{1i} \tag{5.11}$$

and for the 16-year measurement,

$$z_i = \alpha_2 + \beta_2 x_i + \gamma y_i + e_{2i} \tag{5.12}$$

where x, y, z, refer to the seven-, 11- and 16-year measurements respectively. The second equation says that the linear contributions of the 11- and seven-year measurements to the 16-year measurement along paths b and c, add together, although in practice they might combine in some more complicated manner giving rise to interaction terms.

In addition to the problem of measurement error, the e_{1i} and e_{2i} may be correlated so that the separate estimation of these equations will be inefficient. This situation can arise if the equations are incorrectly specified. If this occurs, one approach is to use instrumental variables for the independent variables x_i, y_i, and to obtain the most efficient estimates (although the usual separate equation estimates would be consistent) a "generalised least squares" approach can be used (Zellen, 1962). A special model for dealing with this problem has been developed by Hannan and Young (1977). In practice we can often use transformations to ensure that the basic model fully specifies the relationship between the variables (Goldstein, 1979). In the discussion in the remainder of this section we shall assume that this is the case. Thus the analysis of the first equation can be carried out independently of the second. The same argument applies if further independent variables are added to these equations, each one, except the first, using the previous occasion measurements as the new independent variables. Such a set of equations is known as a "recursive set". When we come to consider the analysis of two or more dependent variables at each occasion, we may no longer have such a recursive set and the analysis will become more complex.

We will see how other independent variables can be introduced into (5.11) and (5.12) using the data from the National Child Development study. The 16-year reading score has been transformed to have a standardized normal distribution and this also gives approximately linear relationships with both seven-year and 11-year scores. It happens to be the case that all three scores can be transformed readily to make all the relationships approximately linear in this example, although in general this may not be possible. The 16-year distribution for given values of seven- and 11-year scores is also approximately normal with approximately constant variance. Further details of the transformations are given in Goldstein (1979). We may take measurement errors into account as before, and taking the reliability of the seven-year score to be 0·79 and that of the 11-year score to be 0·81, we obtain the following estimates of the parameters in equations (5.11) and (5.12).

$$\begin{aligned} y_i &= 0{\cdot}015 + 0{\cdot}841x_i \\ z_i &= 0{\cdot}154 - 0{\cdot}147x_i + 1{\cdot}113y_i. \end{aligned} \tag{5.13}$$

In the second equation we see that the coefficient of x is now negative, having a standard error of 0·051. If we ignore the 11-year measurement and relate z directly to x, we obtain a coefficient of 0·790, and ignoring the seven-year measurement and relating z to y we obtain a coefficient of 0·820. Hence there is a strong positive relationship with both seven- and 11-year scores, but for *given* 11-year score there is a weak negative relationship with seven-year score. This is difficult to interpret and may indicate that non-linear or interaction terms should be included in the model, or perhaps that the change in score between seven and 11 years is more important than the seven-year score itself. However, the addition of non-linear terms does not change this picture to any extent. Nevertheless, the analysis is restricted to the measurements employed, and it might be possible to use other measuring instruments at 11 which would eliminate the seven-year-old contribution.

As before, we can introduce another factor which classifies the children into groups in order to see whether membership of particular groups is associated with change. This time we consider an extension of the social class classification by taking account of social class changes between the three ages. In fact, only 75% are in the same social class at seven, 11 and 16 years so that changes are of practical importance. Figure 5.8 represents the various relationships with 16-year score, this time only including paths relevant to equation (5.12).

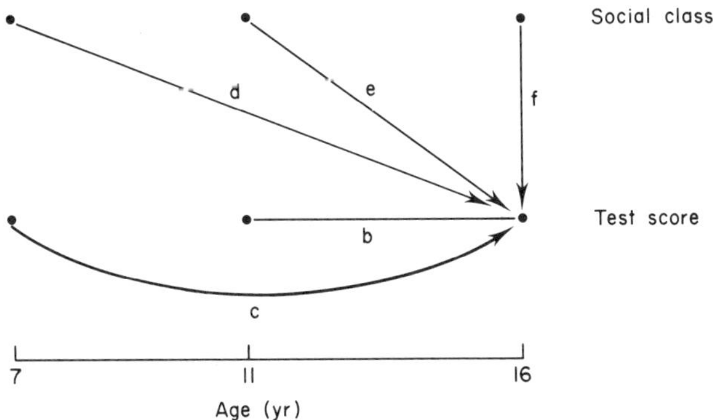

Fig. 5.8 Path diagram for social class and reading scores at seven and 11 years affecting 16 year reading score.

Since we are only interested in effects upon reading scores, possible paths between social classes at each age are omitted. Equation (5.12) has to be extended to allow for social class at the three ages, and one obvious way is as follows:

$$z_i = \alpha + \beta x_i + \gamma y_i + \delta_{1j} + \delta_{2k} + \delta_{3l} + e_{2i} \tag{5.14}$$

where

δ_{1j} refers to categories of seven-year social class
δ_{2k} refers to categories of 11-year social class
δ_{3l} refers to categories of 16-year social class

Among other things, this model implies that the average difference between 16-year social classes is the same for each combination of seven and 11-year scores and social class groups. (The same is true for the 11- and seven-year social class differences.) The alternative, that the difference between the social classes varies with different combinations of social class at seven and 11 years implies that there are interactions present and to incorporate these we need extra terms in (5.14). We considered interactions when we introduced different slopes for the 11 on seven regression lines. Similar problems arise here with scale transformations, and again we may often wish to ignore relatively small interactions for the sake of simple interpretations. Even without interactions, however, there is both a technical and an interpretational problem with (5.14). Despite movements between the social classes, the association between social class at the three ages is high, and attempting to estimate the parameters in (5.14) as it stands would mean that some of the estimates would not be very accurate. Also, since our basic interest lies in changes in social class between ages, it seems more natural to express social class at the three ages in terms of social class at one age and changes from this, thus avoiding also the technical problem. There are three ways of doing this by choosing one of the three ages as the basic one and measuring changes from it. The choice will vary according to the questions the analysis is designed to answer, and we have chosen the age of 11 since it seems to lead to simpler interpretations. Instead of the three factors, $\delta_1, \delta_2, \delta_3$ in equation (5.14) we have three factors as follows:

7 to 11 year change

1. Non-manual → III Manual or IV
2. Non-manual → V
3. III Manual or IV → Non-manual
4. III Manual or IV → V
5. V → Non-manual
6. V → III Manual or IV
7. No change

11 year social class

1. Non-manual
2. III Manual or IV
3. V

11 to 16 year change

1. Non-manual → III Manual or IV
2. Non-manual → V
3. III Manual or IV → Non-manual
4. III Manual or IV → V
5. V → Non-manual
6. V → III Manual or IV
7. No change

Table 5.2 Standardized 16-year reading score related to seven and 11-year reading scores, adjusted for measurement error, social class at 11 and change in social class. (NCDS data.) Figures in brackets are parameter estimates for the model omitting social class changes.

Independent variables		Fitted constant	Standard error
Overall (α)		0·112 (0·115)	
7-year score (β)		0·120 (–0·120)	0·05
11-year score (γ)		1·070 (1·070)	0·06
11-year Social Class	Non-manual (a)	0·039 (0·027)	
	Skilled and semi-skilled manual (b)	0·054 (0·046)	
	Unskilled manual (c)	–0·093 (–0·073)	
7–11-year Social Class change	c → a + b	–0·098	0·043
	b → a	–0·060	0·033
	a + b → c	0·059	0·058
	a → b	0·007	0·041
	No change	0·000	
11–16-year Social Class change	c → a + b	–0·015	0·060
	b → a	0·033	0·034
	a + b → c	–0·087	0·056
	a → b	0·079	0·044
	No change	0·000	
Residual variance = 0·29			

The no-change category of the social class changes is fixed at zero so that the other category constants represent differences from no-change.

Test 7 → 11 social class change after fitting (adjusting for) other factors

χ^2 (4.d.f.) = 10·1*

Test 11 → 16 social class change after fitting (adjusting for) other factors

χ^2 (4.d.f.) = 7·2.

Pairwise contrasts for 11-year social class constants with standard errors in brackets for model omitting social class changes

	b	c
a	–0·019 (0·024)	0·100 (0·039)
b		0·119 (0·031)

Test 11 year social class χ^2 (2.d.f.) = 11·3 **

We note that category 7 comprises all those remaining in each social class. Separate specification of these is clearly unnecessary if 11-year social class is specified. There are now 17 constants associated with social class instead of the nine previously (three at each age). (In fact there are only really 14 and six since we have one side condition for each factor.) What we have done in fact, is to incorporate some of the interaction terms in the specification of the changes, there being 27 ($3 \times 3 \times 3$) possible combinations of social class at the three ages. It would be possible to compare the above groupings of social class with others using a different base age, but for now we shall test whether, when incorporated into a model of the form (5.14), it fits the data well. This may be done by comparing the residual variance when all 27 combinations are fitted with that when only the above 17 are fitted. In fact, there are less than 17, since some of the change categories contain very few cases, and categories 2 and 4 are combined as are categories 5 and 6, leaving 11 constants in all.

A goodness of fit test gives a χ^2 on 16 degrees of freedom equal to 19·5, which is non-significant at the 10% level. (The χ^2 rather than F test is used for simplicity since the degrees of freedom in the denominator of F are very large, see Fogelman and Goldstein, 1976.) Table 5.2 shows the estimates obtained with the model (5.14), taking into account the same reliabilities as before.

There is a suggestion that upward mobility from seven to 11 is associated with a decrease in score when compared with no change. This may be a reflection of the fact that those in the higher social classes at seven and 11 will be expected to score higher at 16 than those in the higher social classes at 11 only. The other contrasts of the change categories with no change do not much exceed their standard error. For this model and the model which omits social class changes, we see that the unskilled manual group at 11 are more than one standard deviation below the other social classes.

5.4 Relationships between Multiple Dependent Variables at Different Occasions

We have, so far, dealt with the case of a single dependent variable at each occasion. In some situations, however, we may have several such variables influencing each other. For example, with two occasions we may have scores both for reading and mathematics as well as, say, social class. The situation in its simplest form can be represented diagrammatically as in Fig. 5.9.

We have mathematics and reading at occasion 1 influencing each other at occasion 2. The double headed arrow at occasion 1 indicates an association between mathematics and reading, which are independent variables at that occasion, and the separate arrows at occasion 2 indicate reciprocal influences for mathematics and reading used as dependent variables. Before commenting on the analysis of this model it is important to be clear about its meaning.

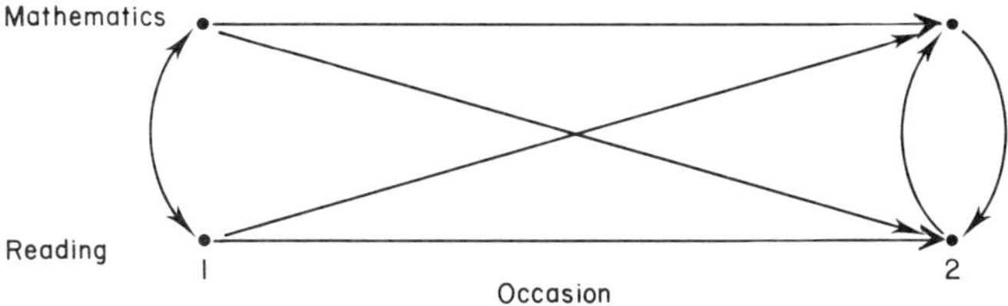

Fig. 5.9 Path diagram for relationships between reading and mathematics at two occasions.

If we consider the occasion 2 reading test score this is related to mathematics and reading at occasion 1 and mathematics at occasion 2. Strictly speaking, it is unlikely that mathematics at occasion 2 has any direct effect on reading at occasion 2 or vice versa, and in general such simultaneous effects will not occur. We would like to have information about these variables between the two occasions in order to study the changing relationships over time. On the other hand, it is reasonable to suppose that reading at occasion 2 also reflects reading attainment during a period prior to occasion 2. If this is true, then a causal influence on mathematics seems more reasonable, although any interpretation of the data is bound to suffer from some vagueness. In some cases we may be able to regard the two occasions as the mid-points of neighbouring time intervals, and the measurements at these occasions to be representative of the values within each interval. If this is extended to several intervals, then it does of course enable us to carry out a detailed time related analysis.

Bearing in mind these reservations, we now write down the equations for this model. Reading at occasion 1 and 2 is denoted by R_1 and R_2 and mathematics by Q_1 and Q_2.

$$\begin{aligned} R_2 &= \alpha_0 + \alpha_1 R_1 + \alpha_2 Q_1 + \alpha_3 Q_2 + e_1 \\ Q_2 &= \beta_0 + \beta_1 Q_1 + \beta_2 R_1 + \beta_3 R_2 + e_2. \end{aligned} \tag{5.15}$$

We assume, as before, that e_1 and e_2 are uncorrelated with each other and the independent variables.

These two equations now constitute a non-recursive system since R_2 and Q_2 appear on both sides of the equations. Thus the ordinary methods of estimation are not applicable. This can easily be seen by substituting for R_2

in the second equation. This gives

$$Q_2 = \frac{1}{(1 - \alpha_3)} [\beta_0 + \alpha_0\beta_3 + (\beta_1 + \alpha_2\beta_3)Q_1 + (\beta_2 + \alpha_1\beta_3)R_1 + (e_2 + \beta_3 e_1)]. \qquad (5.16)$$

Thus Q_2 is not independent of e_1, and hence, as mentioned earlier, the first equation of (5.15) cannot be estimated by ordinary least squares. We can obtain a similar equation to (5.16) for R_2, and this new pair of equations is now recursive since R_2 and Q_2 appear only on the left-hand sides. These are known as the "reduced form" equations. Another problem now arises, however. We see from (5.16) that for the two independent variables R_1, Q_1, the coefficients

$$\frac{1}{(1 - \alpha_3)}(\beta_2 + \alpha_1\beta_3) \text{ and } \frac{1}{(1 - \alpha_3)}(\beta_1 + \alpha_2\beta_3)$$

along with the constant term

$$\frac{1}{(1 - \alpha_3)}(\beta_0 + \alpha_0\beta_3)$$

can all be estimated in the usual way by least squares. Likewise, we can estimate the corresponding three parameters in the equation for R_2. We can also estimate the variances and covariances of the error terms, giving three terms involving the two variances and the parameters β_3 and α_3. Thus we have nine parameter estimates, but we have 10 original parameters in (5.15). Thus an indeterminacy arises, and unless we fix the values of at least one parameter, or specify at least one linear relationship between the parameters, we cannot get unique estimates, and the system is said to be not identified (Johnston, 1972). In an earlier section we gave one method of dealing with the situation where an independent variable was not independent of the error term. This was known as two-stage least squares, where the independent variable is replaced by its predicted value, the predictor variables themselves being uncorrelated with the error terms. In the present case the two predictor variables which satisfy this condition are R_1 and Q_1 and we can write the estimated equation as, say,

$$Q_2 = a + bR_1 + cQ_1. \qquad (5.17)$$

We cannot use the predicted values of Q_2 from this equation in the first equation of (5.15), however, since we are effectively simply using R_1 and Q_1 twice, and we need at least one other predictor of Q_2 before we can use this method. Other similar methods are also available, and the reader is referred to Johnston (1972) for details.

We now give an example of different approaches to this model using the reading data we used before from the National Child Development Study, together with scores on mathematics, both at the ages of seven and 11 years.

In the first case we shall assume that reading affects mathematics at occasion 2 (age 11), but that the path from mathematics to reading is absent. Thus $\alpha_3 = 0$ and (5.15) reduces to two straightforward equations whose coefficients can be estimated separately. Note that, in effect this has made the system identifiable by reducing the number of parameters from 10 to nine. The results are given in Table 5.3. Reliability values of 0·79 for seven-year reading, 0·65 for mathematics at seven, 0·82 for reading at 11 and 0·94 for mathematics at 11 have been used (Goldstein, 1979). In the second case we assume that the cross-lagged pathways are absent, i.e. $\alpha_2 = \beta_2 = 0$ and the variables incluence each other at around 11 years only. The results, using two-stage least squares, are presented in Table 5.4.

Table 5.3 Regressions of Reading and Mathematics at 11 years on Reading and Mathematics at 7 years

Source	Reading		Mathematics	
	Coefficient	Standard Error	Coefficient	Standard Error
Overall	0·018		0·035	
Reading at 7	0·724	0·043	−0·033	0·033
Mathematics at 7	0·170	0·041	0·376	0·022
Reading at 11			0·700	0·040
Residual variance	0·48		0·25	

We now see what happens when two-stage least squares is used to estimate all the parameters in (5.15). We first find the predicted values of reading and mathematics at 11 for each individual, using teachers' ratings at 11 years. After substitution in (5.15) we obtain the results shown in Table (5.5). A comparison of the results in Tables 5.3–5.5 shows the large variation in the coefficient estimates for the different models. For example, when we assume that no path exists from 11-year mathematics to 11-year reading, there is only a small relationship between 11-year mathematics and seven-year reading, which becomes more appreciable when the path is inserted. If we accept the model giving the results in Table 5.5 as the most valid one, then we can conclude that, although reciprocal influence does occur, the influence

Table 5.4 Regression of Reading and Mathematics at 11 and Reading and Mathematics at 7 years with Cross-lagged Parameters set to Zero.

Source	Reading		Mathematics	
	Coefficient	Standard Error	Coefficient	Standard Error
Overall	–0·007		0·035	
Reading at 7	0·374	0·024		
Mathematics at 7			0·363	0·016
Reading at 11			0·682	0·016
Mathematics at 11	0·575	0·018		
Residual variance	0·36		0·25	

Table 5.5 Regressions of Reading and Mathematics at 11 years on Reading and Mathematics at 7 years using Two-stage Least Squares

Source	Reading		Mathematics	
	Coefficient	Standard Error	Coefficient	Standard Error
Overall	0·007		0·036	
Reading at 7	0·600	0·023	0·261	0·024
Mathematics at 7	0·071	0·024	0·487	0·026
Reading at 11			0·395	0·016
Mathematics at 11	0·331	0·015		
Residual variance	0·45		0·35	

of reading on maths is stronger than vice versa, and in particular that none of the possible pathways should be assumed to have zero or near-zero coefficients. A further question is whether the assumptions behind the two-stage least squares technique are correct. It may well be that teachers' ratings are, in fact, correlated with the measurement error. There is also the question of how adequate are the reliability estimates, and a further discussion of these problems can be found in Goldstein (1979).

We can introduce further independent variables into this system as before in order to study, for example, social class or family size effects. The analysis of such models is obtained by a straightforward extension of the analysis we have already discussed, and we shall not pursue any details. When we consider models relating to three or more occasions we may also encounter measurement errors in the independent variables as well as non-recursive systems, and these can be dealt with by straightforward extensions of the above methods, either by inserting reliability estimates or using two-stage least squares for these independent variables.

5.5 More General Models for Relating Measurement at Different Occasions

We now outline ways in which the models so far discussed can be extended, with a brief discussion of the attendant problems of analysis and interpretation. The statistical models arise largely from the general model of Joreskog (1970) and the reader is referred to that paper for more detailed results.

Reverting to equations (5.11) and (5.12), let us write down a more general structure for these, namely

$$\begin{aligned} y_i &= \alpha_1 + \beta_1 x_i + e_{1i} \\ z_i &= \alpha_2 + \beta_2 x_i + \gamma y_i + e_{2i} \end{aligned} \tag{5.18}$$

and

$$\begin{aligned} x_i &= X_i + u_{1i} \\ y_i &= Y_i + u_{2i} \\ z_i &= Z_i + u_{3i} \end{aligned} \tag{5.19}$$

where u_1, u_2, u_3 are measurement errors.

If we count the number of parameters involved here, including the variances and covariances of X, Y, Z the variances of u_1, u_2, u_3, e_1, e_2 and the coefficients, there are 16 in all, or 17 if we allow e_1 and e_2 to be correlated. However, we have only nine variances, covariances and means available and so the model is not identified, as we saw with (5.15). There are several ways of resolving the problem.

First, we can predetermine the values of some of the parameters or specify linear relationships between them as we did before. We could, for example, specify the values of the variances of the measurement errors giving us our original model with known reliabilities. An alternative is to attempt to take further measurements which are related to X, Y, Z in order to help to identify their variances. Suppose we have alternative tests x^1, y^1, z^1 such that

$$\begin{aligned} x_i^1 &= aX_i + u_{1i}^1 \\ y_i^1 &= bY_i + u_{2i}^1 \\ z_i^1 &= cZ_i + u_{3i}^1. \end{aligned} \tag{5.20}$$

We have introduced six extra parameters, but 18 extra means, variances and covariances between the two sets of measurements, so we now have 27 means, variances and covariances in order to estimate 23 parameters, and the model is identified. In fact equations (5.18–5.20) taken together represent a particular kind of factor analysis model with one common factor for each pair of observed variables. More general factor analysis models can be specified in a similar way, giving longitudinal factor analysis models, where relationships across occasions are specified between the unobserved factors, rather than the observed measurements. The reader is referred to Joreskog and Sorbom (1976) for details and examples. Another approach, with models using unreliable measurements at several occasions, is given by Kadane *et al.* (1976), who carry out a detailed analysis of a four-occasion model of IQ change, introducing measurement errors without assuming full multivariate normality and using two-stage least squares and a generalization known as three-stage least squares.

By including further independent variables into these models and specifying simple functions of them, we can obtain more means, variances and covariances in order to exceed the number of parameters to be estimated. Sorbom (1976) gives an example where relationships are compared within several groups of individuals, that is using a categorical independent variable which divides individuals into groups. These models, however, assume that the variables have multivariate normal distributions, and there appears to be little information on how reasonable this assumption is in practice, and how departures from it can affect the analysis. Furthermore, rather similar criticisms can be levelled against the use of these complex models as can be made in general against Factor Analytic or Latent Structure models in general, namely that the necessity for specifying the values of certain parameters can lead to arbitrariness, with very different models appearing to fit the same set of data quite adequately. Also, these models assume that parameters either have fixed values or are to be estimated from the data. In practice, as we have seen, we may have estimates of, for example, reliabilities which have a sampling variability. In the models of the previous section this can be taken into account, its effect being to increase the standard errors of the parameter estimates. There appears to be no really satisfactory method of dealing with such sampling variability in the models of the type proposed by Joreskog (1970), which seems to be a further reason for treating them with some caution.

5.6 Models for Discrete Data

The models we have considered have assumed that the response or dependent variable is continuous, that is, measured on an interval or ratio scale, and the equations have been expressed in terms of the mean or expected values of these variables. We now look at models for variables measured on qualitative or categorical scales without any necessary ordering, although, as we shall see, we can also carry out some analyses of ordered categories using these models.

The simplest case of a qualitative variable is one with just two response categories, for example social class divided into non-manual and manual groups, or an attainment, categorized as "average or above" and "below average". Interest now focuses on the probability, or expected proportion, in each category. Since the categories we shall deal with are mutually exclusive, in the case of just two categories we are only interested in one of the probabilities since the other is simply one minus this probability.

To introduce the basic model we shall consider a very simple example using the NCDS data. The reading scores at 16 have been categorized into two groups, and the 11-year scores into five groups.

Table 5.6 Poor Reading at 16 by 11-year Reading Score (NCDS Data) (percentages in brackets)

Score in 16-year reading test \ Score in 11-year reading test	Number of children in each score range					
	0–5	6–12	13–19	20–26	27–35	Total
0–11 (poor readers)	151 (51·7)	170 (8·6)	12 (0·4)	4 (0·2)	1 (0·2)	338 (4·1)
12–35	141 (48·3)	1818 (91·4)	3321 (99·6)	2170 (99·8)	470 (99·8)	7920 (95·9)
Total	292 (100)	1988 (100)	3333 (100)	2174 (100)	471 (100)	8258 (100)
Predicted % of poor readers (logit model)	51·7	8·3	0·6	0·05	0·002	

Table 5.6 presents the numbers of children, together with the observed percentages in the two 16-year data categories, for each of the 11-year categories. We see immediately that there is a strong association between reading at 11 and 16, over half of those in the lowest category at 11 also being in the lowest (poor reading) category at 16, as compared with an overall 4·1% in this category at 16. Note particularly that the choice of categories is

essentially arbitrary; here, for example, about 4% are in the lowest category at each age. The same situation occurs if we wish to relate percentages in categories to age, although here we have the additional problem (encountered earlier) of using different measurements at each age. On the other hand, there are variables such as social class, where we can retain the same categorization of the same variable at each occasion.

A simple model for Table 5.6 is as follows:

$$p_j = a + b_j \tag{5.21}$$

where p_j is the expected proportion of 16-year-old poor readers in the jth 11-year category, and the observed proportion, in a random sample, has a binomial distribution about this mean value. This is analogous to the ordinary one-way analysis of a variance model. Alternatively, we could attempt to summarize the relationship by giving scores x_i to each category of the 11-year score (for example the mid-point value of the categories), so that (5.21) becomes

$$p_i = a + bx_i \tag{5.22}$$

which is analogous to ordinary simple linear regression. In (5.21) the parameters a and b_j can be estimated respectively as the overall and the deviation from overall, proportions. The efficient estimation of the parameters in (5.22) involves a weighted linear regression analysis (see e.g. Armitage, 1955). Using the mid-point of the 11-year reading categories, equation (5.22) gives $a = 2{\cdot}04$ and $b = -0{\cdot}08$. This implies, that for an 11-year reading score greater than 26, the expected proportion is negative! Even if we fit an extra quadratic term the same occurs for 11-year scores between 17 and 32. In fact, although by definition p_i can never be negative, there is no restriction in equation (5.22) which actually prevents this occurring, with the results we have seen. This will usually be a problem when small proportions are involved and when there are several ways of classification. We could continue to fit higher order terms, but then we lose the simple summary provided by an equation such as (5.22).

We look, therefore, for a transformation of p_i which will provide a simple summary of the data and guarantee non-negative predicted values. One possible candidate might be log p_i which ranges from $-\infty$ to 0 as p_i goes from 0 to 1. In fact, we choose log $[p_i/(1 - p_i)]$ since this always guarantees that p_i lies between 0 and 1 and is symmetrical in p_i and $q_i = (1 - p_i)$, and hence it does not matter which of the two dependent variable categories we choose to work with. We rewrite (5.22) as

$$\log\,[p_i/(1 - p_i)] = \operatorname{logit} p_i = c + dx_i. \tag{5.23}$$

The bottom line of Table 5.6 gives the predicted percentage of poor readers using this model, and we see that apart from the two categories with very small numbers the predicted percentages agree closely with the observed percentages. We pursue this type of transformation in the next section where we give a simple outline of the "log-linear" model, and show how it can be used to write down models analogous to the more familiar ones for continuous dependent variables. A fuller treatment of long-linear models can be found in Bishop *et al.* (1975). We then show how path analysis models can be constructed and how "growth curve" type models can be developed. Although the log transformation will be used mainly, there will be occasions when the untransformed proportions are useful. Finally, we shall consider briefly the special case of the analysis of ordered dependent variables, and the problem of measurement errors.

5.6.1 Log-linear Models

Consider again Table 5.6 and write the observed proportion in cell (i,j), where i refers to the rows and j to the columns, as

$$p_{ij} = r_{ij}/N$$

where r_{ij} is the frequency in cell (i,j) and N the total number in the table. If we denote by π_{ij} the expected or population value of p_{ij}, we can write down a model analogous to the usual 2-way analysis of variance model, namely

$$\log \pi_{ij} = a + b_i + c_j + d_{ij}, \qquad i = 1, \ldots r, \quad j = 1, \ldots c \qquad (5.24)$$

with $\sum_i b_i = \sum_j c_j = \sum_j d_{ij} = \sum_i d_{ij} = 0$.

This model reproduces the observed proportions. If the d_{ij} terms are zero then the cell probabilities depend only on constants fitted to the row and column margins, and this is simply the case of no association between rows and columns of the contingency table. Testing the hypothesis $d_{ij} = 0$ is hence a test of the association in a two-way contingency table.

In our example we have $r = 2$, $c = 5$. We can write

$$\log \pi_{1j} - \log \pi_{2j} = (b_1 - b_2) + (d_{1j} - d_{2j}).$$

Now $\log \pi_{1j} - \log \pi_{2j} = \log \pi_{1j}/\pi_{2j} = \text{logit}\, \pi_{1j} = \text{logit}\, \pi_j$ say, and if we write $b = b_1 - b_2$, $d_j = d_{1j} - d_{2j}$, we have

$$\text{logit}\, \pi_j = b + d_j \qquad (5.25)$$

which is equation (5.21) with a logit transformation. Alternatively we can write (5.25) as

$$\pi_{ij} = \frac{\exp(b + d_j)}{1 + \exp(b + d_j)}.$$

Consider now the general two-way table model (5.24). We have

$$\pi_{ij} = \exp(a + c_j + b_i + d_{ij}).$$

We wish now to treat the set of row probabilities for each column as the dependent variable vector with r elements, each corresponding to a row category. Since an individual must be in one of these row categories, we have the constraint

$$\sum_i \pi_{ij} = 1 \tag{5.26}$$

which gives $$\exp(a + c_j) = \frac{1}{\sum_i \exp(b_i + d_{ij})}.$$

Hence we can write $$\pi_{ij} = \frac{\exp(b_i + d_{ij})}{\sum_i \exp(b_i + d_{ij})}. \tag{5.27}$$

This is a multivariate generalization of the ordinary logit given above.

We can, of course, extend (5.24) to include further variables by defining three-way or higher order tables, and (5.27) can be extended so that the π_{ij} refer to the probabilities in a two-way or higher order table considered as a set of dependent or response variables. These response variables may also have a particular structure, for example being linearly related to marginal totals. The reader is referred to Bock (1976a) for details of such models. In the following examples we shall discuss mainly the ordinary logit model, the multivariate extension being straightforward. We note, in particular, that (5.27), with (5.25) being a special case, can be analysed by any method which can analyse log-linear models of the kind (5.23), with the additional constraint (5.26). Computer programs exist which will, with careful specification, analyse models of this type (see for example Nelder, 1974, Baker and Nelder, 1978). The results of such an analysis will provide parameter estimates, standard errors, and a goodness of fit test. The latter provides a means whereby we can judge whether the predicted cell frequencies for the original contingency table are significantly different from those observed, i.e. whether the model fits the data. Different models can be tested in this way and also against one another. As in ordinary analysis of variance, we can test subsets of parameters within an overall fitted model by studying differences in χ^2 statistics for the fitted models, as we illustrate below.

5.6.2 Log-linear Path Analysis Models: Recursive Systems

The path analysis model for just two occasions used the data in Table 5.6. To illustrate the general approach for categorical data we shall use an extension of Table 5.6 by incorporating reading score at seven years and social class at 11 years. Each variable will be grouped into only two categories in order to simplify the presentation, but the extension to more than two categories is straightforward, using the approach of the previous section.

Table 5.7 Poor reading at 7 by poor reading at 11 by poor reading at 16 by Social Class at 11. Expected frequencies using (5.28) and (5.29) and 7 × 11 interaction are given in brackets. (NCDS data.)

		Social Class			
		Non-manual 16-year score		Manual 16-year score	
7-year Score	11-year Score	0–11	12–35	0–11	12–35
0–8	0–5	6 (1·6)	3 (7·4)	55 (60·9)	29 (23·1)
	6–35	2 (1·5)	23 (23·5)	43 (42·9)	156 (156·1)
9–30	0–5	7 (5·5)	23 (24·5)	83 (83·2)	86 (85·8)
	6–35	7 (13·0)	2615 (2609·0)	135 (129·1)	4987 (4992·9)

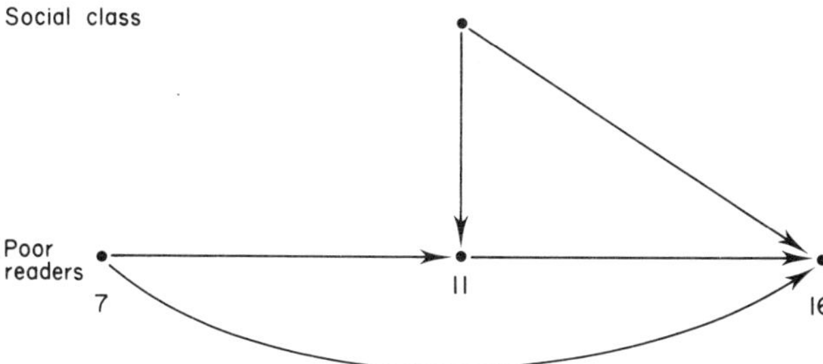

Fig. 5.10 Path diagram for effect of social class on proportion of poor reading scores at seven, 11 and 16 years.

We can write down the two equations represented by Fig. 5.10 thus:

$$\text{logit } y_{ij} = a_1 + b_{1i} + c_{1j} \tag{5.28}$$

$$\text{logit } z_{ijk} = a_2 + b_{2i} + c_{2j} + d_{2k} \tag{5.29}$$

where subscript i = 1,2 refers to the seven year reading category
j = 1,2 refers to the eleven year social class category
k = 1,2 refers to the eleven year reading category
y,z are the probabilities of being a poor reader at 11 and 16 years

with $\sum_{i=1}^{2} b_{1i} = \sum_{j=1}^{2} c_{1j} = 0$, etc.

We see that this is a recursive system, as it was defined for continuous models. Table 5.7 is, in effect, a four-way contingency table with each factor having two categories, i.e. a 2^4 table. Equation (5.28) is concerned solely with the three-way table formed by collapsing over the 16-year reading score. Equation (5.29) uses this three-way table in order to define the independent variables. We therefore analyse (5.28) and (5.29) separately to obtain the parameter estimates.

Table 5.8 Maximum Likelihood Parameter Estimates for Equations (5.28) and (5.29) (parameter g refers to the 7 × 11 interaction as do the figures in brackets)

Parameter	Estimate	Standard Error
a_1	−2·57	0·10
b_{11}	1·32	0·07
c_{11}	−0·46	0·09
a_2	−1·84 (−1·84)	0·14 (0·13)
b_{21}	0·94 (0·80)	0·08 (0·08)
c_{21}	−0·86 (−0·82)	0·12 (0·12)
d_{21}	1·65 (1·44)	0·08 (0·08)
g	(0·40)	(0·08)

Overall Goodness of fit χ^2(5d.f.) = 36·0***

Goodness of fit eqn. (5.28) χ^2(1d.f.) = 3·6

Goodness of fit eqn. (5.29) χ^2(4d.f.) = 32·4***

For eqn. (5.29) fitting 7 × 11 interaction gives goodness of fit χ^2(3d.f.) = 10·7*; fitting 7 × social class interaction gives goodness of fit χ^2(3d.f.) = 28·8***; fitting 11 × social class interaction gives goodness of fit χ^2(3d.f.) = 25·2***.

Table 5.8 gives the estimates for (5.28) and (5.29) together with their standard errors. Equation (5.28) which is an additive "main effects" logit model, fits the data quite well, but we require a term for the interaction between seven and 11 year scores to be added to (5.29). Even the model including interaction, however, gives a significant goodness fit test statistic. In addition to estimating predicted proportions we may wish to find the expected frequencies in Table 5.7. If we had only equation (5.29) we would do this simply by applying the predicted probabilities to the frequencies in the three-way table which defines the independent variables. Using (5.28) in addition, however, we can predict the frequencies in the three-way table from the two-way table defined by social class and seven-year reading category. Thus the pair of equations gives a set of expected frequencies estimated as follows:

$$F_{ijkl} = F^1_{ijl} F^2_{ijkl}/n_{ijl}$$

where F^1_{ijl} is the expected frequency and n_{ijl} the observed frequency in the lth 11-year reading category using equation (5.28), and F^2_{ijkl} the corresponding expected frequency using (5.29) (see Goodman, 1973). These estimated frequencies are given in brackets in Table 5.7.

5.6.3 Log-linear Path Analysis Models: Non-recursive Systems

As with continuous variables, we may have path models containing reciprocal influences. We recall that we could obtain a reduced form for equations such as (5.15), which constituted a recursive system, although the individual parameters could not necessarily be identified. The general method of solution proposed was to use two stage least squares, and introducing suitable instrumental variables.

In the present case, we have seen that our logit model equations are derived from log-linear models containing main effects and interaction terms. For models based on three-way or higher order tables some of the high order interaction terms will typically be zero or near zero. If the equations in a system based on a given table are to be satisfied simultaneously, we can obtain the equivalent of the reduced form as follows. We specify a new log-linear model which is simply that one which has zero interactions wherever any interaction in the separate equations is zero. The new log-linear model will thus be consistent with the set of simultaneous equations and we can estimate its parameters and hence identify and estimate the parameters of derived logit equations. We shall present a simple example of such a model to illustrate the procedure. More complex examples and further details can be found in Goodman (1973).

Table 5.9 Social Class at 11 years by poor reading and poor mathematics test scores (NCDS data)

Reading Score \ Mathematics Score	Non-manual 0–1	Non-manual 2–40	Manual 0–1	Manual 2–40
0–5	27	43	220	225
6–35	22	3886	181	7336

Estimated parameters for equations (5.30) (5.31). Note $b_{11} = -b_{12}$; $b_{21} = -b_{22}$; $c_{11} = -c_{12}$; $d_{21} = -d_{22}$.

Parameter	a_1	b_{11}	c_{11}	a_2	b_{21}	d_{21}
Estimate	−2·40	−0·54	1·91	−2·02	−0·40	1·91
Standard Error	0·08	0·08	0·06	0·07	0·07	0·06

Test for goodness of fit for (5.30) or (5.31) χ^2 (1d.f.) = 8·9**
Estimating parameters for equation (5.31) setting $d_{21} = 0$

	a_2	b_{21}
parameter	–3·42	–0·60
standard error	0·07	0·07

Table 5.9 is a three-way contingency table of social class at 11 years by reading at 11 years by mathematics at 11 years, each dichotomized into two groups. Figure 5.11 shows a simple path diagram with mathematics and reading influencing each other.

The logit equations are,

mathematics: $$\text{logit}\ \pi_{ij} = a_1 + b_{1i} + c_{1j} \qquad (5.30)$$

reading: $$\text{logit}\,\pi_{ik} = a_2 + b_{2i} + d_{2k} \tag{5.31}$$

where $i = 1, 2$ refers to social class, $j = 1, 2$ to reading and $k = 1, 2$ to mathematics.

In equation (5.30) the (i,j) interaction is zero and in (5.31) the (i,k) interaction is zero. For both the corresponding log-linear models this implies that the ijk interaction is zero and hence the required log-linear model is as follows,

$$\log \pi_{ijk} = A + B_i + C_j + D_k + (BC)_{ij} + (BD)_{ik} + (CD)_{jk}. \tag{5.32}$$

The parameter estimates for (5.30) and (5.31) are given in Table 5.9. We see that the parameters for the two equations are similar (with $c_{11} = d_{21}$ since both are derived from (CD) in (5.32)). We note that the goodness of fit test gives a significant result, suggesting that we cannot produce additive logit models for these data in the form (5.30) and (5.31).

If we delete the arrow from C to B in Fig. 5.11, on the grounds that reading is not influenced by mathematics, then we are back with a recursive system, the parameter estimates for which are also given in Table 5.9. It should be noted particularly that, unlike the continuous variable case, a general feature of these non-recursive systems is that estimates for reciprocal paths, such as those in Fig. 5.11, are constrained to be equal.

More complex path models can often be split into separate components, each related to a table of a different order and in which lower order tables form the margins of higher order tables and can therefore be estimated separately, as with fully recursive systems.

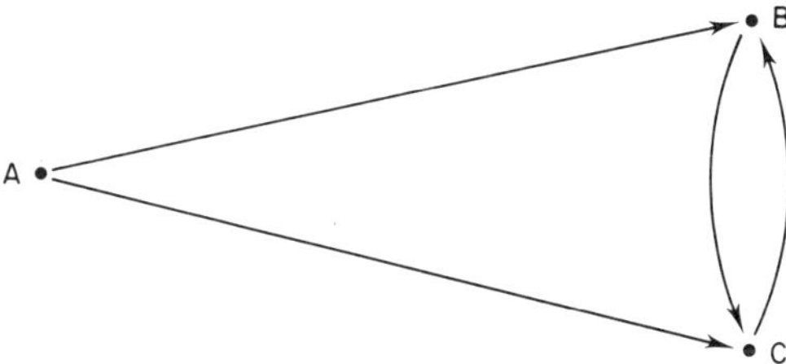

Fig. 5.11 Path diagram for effect of social class at 11 years (A) on proportion of poor reading scores at 11 years (B) and proportion of poor mathematics scores at 11 years (C).

5.6.4 Growth Curve Type Models

Table 5.10 gives the numbers of children in non-manual and manual social class groups in a three-way table using the occasions of birth, seven and 11 years.

Table 5.10 Social Class at birth by Social Class at 7 by Social Class at 11 (NCDS data)

	Social Class at 11			
	Non-manual		Manual	
Social Class	Social Class at 7		Social Class at 7	
at birth	Non-manual	Manual	Non-manual	Manual
Non-manual	2150	156	118	366
Manual	679	549	260	5692

Suppose that we wish to study how the proportion in these two social class groups changes with age, this being the analogue of the growth curve model where we use the same measurement at each occasion, and relate it to age or time.

We have the equation

$$\pi_{ij} = a + b_{ij} \tag{5.33}$$

or its log form

$$\log \pi_{ij} = c + d_{ij} \tag{5.34}$$

where $j = 1, 2, 3$ refers to age, and $i = 1, 2$ refers to social class. If we assume a linear relationship with age, then, e.g. (5.34), becomes

$$\log \pi_{ij} = c + d_i x_j \tag{5.35}$$

where x_j is the age at occasion j.

The dependent variables in these models are based on the separate marginal proportions at each age, rather than the individual cell proportions of the three-way table, so that we cannot use directly the methods of analysis described in the previous section.

To understand how the analysis proceeds, we consider first the case where different samples are available at each occasion. We now have a simple 2×3 table which we can analyse using a model such as (5.24). When the data are, in fact, longitudinal, and the samples at each occasion are not therefore independent, we need to take account of the dependency in the analysis, although the structure of the models and their interpretations are analogous to those we would use in the separate sample case. Full details of the analysis are given in Appendix 5.1, and it can be thought of in three stages, which we shall illustrate with the data in Table 5.10.

First, we specify the marginal probabilities of interest by suitable combinations of the proportions in the original table. These are given in Table 5.11. We then set up a model, such as (5.35) relating these to a function of the occasion ages, and possibly other independent variables. We might for example add another way of classification by introducing sex as a variable, giving an overall four-way table and adding a further term to (5.35). Finally we estimate the parameters of the model, and test hypotheses about them, including an overall test of goodness of fit. Table 5.11 gives the predicted

Table 5.11 Models Fitted to Data in Table 5.10 Relating Proportion in the Non-manual Group to Age

Age	Observed proportions	Estimated proportions		
		Linear model	Log-linear model	Logit-linear model
0	0·281	0·280	0·280	0·280
7	0·322	0·326	0·325	0·326
11	0·355	0·353	0·354	0·354
χ^2 goodness of fit (1.d.f.)		4·5*	1·9	2·9

proportions using the linear log-linear (5.34) and a logit linear model, with age as the independent variable. All three models give a reasonably close fit to the observed proportions, although the log-linear model seems slightly better than the other two. For this age range any of the models would give similar predictions, although care should be used in extrapolating the prediction equations to later ages.

5.6.5 Ordered Dependent Variables

Where the categories of the dependent variables are ordered, we may wish to assign a score to each category and relate the mean score to independent variables. This can be done using the model described in Appendix 5.1. It is also worth noting that, where there is just one (dependent variable) occasion, we can use this procedure to relate such a mean score to a function of independent variables which, in particular, might be those measured at previous occasions. Hence we can specify a path analysis type model, and, at least for recursive systems, the analysis proceeds along similar lines to those already discussed.

In Chapter 4 we described the non-parametric analysis of the growth curve model where the dependent variable was ranked. As with categorized data, if we consider the case of only one occasion, then we can relate the mean rank to a function of independent variables which might themselves consist of ranked data, for example involving polynomial functions of these ranks. In particular, they may be variables at one or more previous occasions, and so we obtain path analysis type models.

5.6.6 Measurement Error

There seems to be no simple theory to deal with measurement errors for discrete data. Where individual responses are classified into categories there are many different ways in which misclassifications can occur, each of which will require a different model to describe it. For example, one simple model specifies that there are a set of "true" proportions in each category, and where a misclassification occurs, its probability of being in another given category is proportional to the true proportion of individuals in that category (Battese and Fuller, 1973). Mote and Anderson (1965) consider one model where the misclassified responses occur with equal probability in each of the other categories, and another model where misclassified responses fall only into "adjacent" categories. In general, where there are measurement errors, the observed proportions are not consistent estimators of the true population proportions, although for some models, for example that of Battese and Fuller (1973), the estimators are consistent.

There seems to be little general guidance which can be given where discrete data are subject to measurement error. It may sometimes be possible to estimate the errors by, for example, remeasuring individuals, and there may occasionally be theoretical grounds for assuming a particular type of error model. If errors occur as a result of coding or data transcribing, then they can often be estimated, and usually considerably reduced if care is taken (see Chapter 1).

5.6.7 Further Considerations

We conclude this section on categorized data by drawing attention to some developments of the above models.

Koch *et al.* (1972) discuss the situation where data are missing from a multiway contingency table and give methods for testing some aspects of the representativeness of the missing data and, assuming elements missing at random, show how all the available information can be used to provide efficient estimates. Goodman (1974) discusses latent structure models for categorical data, where the observed variables are expressed as functions of fewer unobserved latent variables. Some of these models are analogous to those mentioned in Section 5.5.

5.7 Event Data

The type of study considered in previous sections is based on a set of well defined occasions. Individuals are measured at these occasions and the measurements provide a description of time-based relationships. An alternative way of following an individual through time is to record the times at which specified events take place. Such studies are found, for example, in epidemiology and clinical trials where the survival time of an individual is of interest, or in physical growth where we may be interested in the age at which menarche occurs. Studies of careers may also have such a structure, where the events being studied are changes of job. Although it is not one of the aims of this volume to deal with such data in any detail, a brief introduction to their analysis will be given, and the reader is referred to Hills (1974) for a more extensive introductory account and also to Armitage (1971).

The basic concept is that of a hazard. This is the probability that, given that an event has not occurred before a time or age t, it will happen in the next unit of time. In general this hazard will depend on time and is denoted by $\lambda(t)$, although in some situations $\lambda(t)$ can be regarded as constant over a time interval. In this case we have what is known as a "poisson process", where the probability that an event does not occur by time t, for example that an individual has survived until this time, is given by

$$e^{-\lambda t}. \tag{5.36}$$

In general, for a sample, we can estimate $\lambda(t)$ at several different values of t, either by observing the times at which events occur or by observing how many events occur in narrow intervals of time. In the latter case, if we assume that $\lambda(t)$ remains constant within these time intervals, then the probability that an event does not occur for each of these intervals is given by (5.36). If the intervals are small enough, then we can estimate (5.36) by $\frac{d}{n}$ where d is the number of events occurring in the interval and n is the number of individuals for whom the event had not occurred by the start of the interval. If there are k intervals up to time T then the probability that an event does not occur before time T is given by,

$$F(T) = (1 - \frac{d_1}{n_1})(1 - \frac{d_2}{n_2}) \ldots \times \ldots (1 - \frac{d_k}{n_k}) \tag{5.37}$$

or assuming (5.36) within each interval,

$$F(T) = \exp\left(-\sum_{i=1}^{k} \lambda_i t_i\right). \tag{5.38}$$

$F(T)$ is known as the "survivor function", the terminology deriving from mortality studies. For a sample, $F(T)$ can be estimated in the above way, or alternatively we may specify a particular relationship with time, of which (5.36) is the simplest. For example, when studying the distribution of the age at menarche the following relationship is found to be a good approximation:

$$\log(F/(1-F)) = a + bt. \tag{5.39}$$

We have assumed that the times of occurrence of all events are known, at least within a narrow interval. In practice, however, this will not be true generally. We may have information which tells us only whether or not the event has occurred before a particular time. In fact, this latter case is a common situation which can arise when individuals are questioned at given occasions, and is often to be preferred over attempts to obtain the actual time of occurrence by getting individuals to recall the precise time of the event. For methods of analysing mixtures of these two types of data the reader is referred to Swan (1969), Peto (1973) and Atwood and Taube (1976).

A different approach to the analysis of survival time distributions studies the way in which $\lambda(t)$ depends on a set of independent variables, for example different groups of patients in a clinical trial. A basic model can be written,

$$\lambda(t) = \lambda_0(t)\exp\sum_i \beta_i x_i \tag{5.40}$$

where $\lambda_0(t)$ is treated as a "nuisance" variable and interest centres on estimating the coefficients β_i. For example, if there is just one x variable which is a dummy variable defining two groups, then the ratio of the hazards for the two groups depends on only one parameter and is simply e^{β}. Cox (1972) develops this approach and the reader is also referred to Clayton (1978).

5.8 Retrospective Data

Throughout this chapter we have assumed that a value referring to a specific occasion is actually obtained from a measurement made at that occasion. Nevertheless, it is quite possible, and indeed very common, to attempt to obtain such a value by using measurements made at other, subsequent occasions. For example, we might ask a subject how many cigarettes he or she smoked three months previously or we might ask a child's schoolteacher to recall his performance a year previously. Such data are called retrospective. The term is applied also to data obtained from archives or records.

If, by retrospective questioning for example, we obtain data relating to a set of occasions, we can in principle analyse them using the models discussed in this chapter. The two problems of bias and unreliability arising from faulty recall, misunderstanding, etc., however, will modify both the analysis and its interpretation. Discussions of the problems which can arise are given in Moss and Goldstein (1979). Where problems of bias are absent, we may be able to think of an extra measurement error component being added as a result of, say, random recall error and this will be incorporated into any existing measurement error. So long as an estimate of the magnitude of the error is available, the techniques described earlier can be used. When fitting growth curve type models, and where the measurements have different reliabilities at each occasion, we may take account of this by weighting the measurements at each occasion in proportion to their reliabilities.

It is often assumed that retrospective data are rather poor substitutes for prospective or fully longitudinal data. There are situations, however, when retrospectively collected data provide the most efficient method of carrying out a study. For example, if we are studying the birth factors influencing the occurrence of a rare childhood disease, then it may be very inefficient to follow up a very large sample of births in order to guarantee a sufficient number of cases who will acquire the disease. An alternative strategy would be to sample those cases who had acquired the disease, together with a sample of controls who had not, and then obtain the birth information by methods such as retrospective questioning and use of records. The problems of reliability and bias may still arise of course, and we also have the problem of obtaining representative samples of cases and controls. Nevertheless, these advantages may be offset by the increased efficiency and reduced cost of such a study. Models for the analysis of such studies have generally been concerned with the risks of developing diseases and the comparison of these risks over the categories of other factors. Because of the way in which cases and controls are selected, no estimate of overall risk can be obtained directly, but the "relative risk" or the ratio of risks for two categories or groups of individuals can be estimated. These can then be treated as dependent variables in a model with other independent factors, using for example logit-linear models of the kind discussed in Section 5.6.1. Such models are discussed by Prentice (1976), and the reader is also referred to the paper by Cornfield and Haenszel (1960). It has even been suggested (Mantel, 1973) that large prospective studies can be treated as if they were retrospective by taking all individuals with, say, a disease, and sampling the controls. The loss in precision is often small when this is done, and the saving in analysis time can be considerable. Of course, in this case, an estimate of overall risk is also available.

In many applications it is unrealistic to suppose that cases can be sampled at a fixed occasion, and we may need to consider the incidence over a period of time. Thus we would be dealing with event data of the type discussed in the previous section, where the hazard is now the incidence rate and models such as (5.40) can be employed. A discussion of such models is given by Clayton (1979), and the reader is also referred to Liddell *et al.* (1977).

Appendix 5.1

A Model for the Analysis of Longitudinal Categorical Data

The following outline summarizes the analysis presented by Koch *et al.* (1977) and uses mainly their notation.

Let $g = 1, \ldots, d$ index the d occasions. At each occasion there is a response vector with L categories, hence there are $r = L^d$ possible response configurations, i.e. cells of the multiway table in which each way of classification is a response vector at one occasion.

Let $j = 1, \ldots, r$ index these cells;

let $i = 1, \ldots, s$ index s subgroups which may represent, for example, a cross-classification of independent variables.

Denote by p_i the $(r \times 1)$ vector of observed proportions from the ith subgroup out of the total for that group (n_i).

Denote by P the $(sr \times 1)$ vector formed by combining the p_i

$$P^{\mathrm{T}} = (p_1^{\mathrm{T}}, \ldots, p_s^{\mathrm{T}})$$

P is the unrestricted maximum likelihood estimator of π where

$$\pi^{\mathrm{T}} = (\pi_1^{\mathrm{T}}, \ldots, \pi_s^{\mathrm{T}}).$$

The covariance matrix of P is block-diagonal and a consistent estimator for the ith diagonal matrix is given by the usual multinomial formula,

$$V_i(p_i) = \frac{1}{n_i}[Dp_i - P_i P_i^{\mathrm{T}}]$$

Where Dp_i is an $(r \times r)$ diagonal matrix with the elements of p_i on the main diagonal.

In the example in the text we have $r = 2 \times 2 \times 2 = 8$ and $s = 1$. Hence $P^{\mathrm{T}} = p_1^{\mathrm{T}} = (0{\cdot}216, 0{\cdot}012, 0{\cdot}016, 0{\cdot}037, 0{\cdot}068, 0{\cdot}026, 0{\cdot}055, 0{\cdot}571)$ where the components of P are ordered 111 112 121, etc.

We now form the response functions, i.e. the proportions at each occasion.

Define the $(u \times sr)$ matrix A_1 which will be of the block diagonal form $A_1 = A \otimes I_s$ where we assume the same response functions in each group. In the present case we have

$$A_1 = A = \begin{bmatrix} 1 & 1 & 1 & 1 & 0 & 0 & 0 & 0 \\ 0 & 0 & 0 & 0 & 1 & 1 & 1 & 1 \\ 1 & 1 & 0 & 0 & 1 & 1 & 0 & 0 \\ 0 & 0 & 1 & 1 & 0 & 0 & 1 & 1 \\ 1 & 0 & 1 & 0 & 1 & 0 & 1 & 0 \\ 0 & 1 & 0 & 1 & 0 & 1 & 0 & 1 \end{bmatrix}.$$

If we write $G = AP$ then the first element of the $(u \times 1)$ vector G is the proportion of non-manual at birth, the second element is the proportion of manual at birth, etc. Thus the rows of A define the response proportions at each occasion. If we use G as the response vector, then the model will be a linear one in the probabilities. We may, however, wish to use a model which is linear in the logarithms in which case we can define the vector

$$\log G$$

or possibly

$$\operatorname{logit} G = A_2 \log G$$

where $$A_2 = \begin{bmatrix} 1 & -1 & 0 & 0 & 0 & 0 \\ 0 & 0 & 1 & -1 & 0 & 0 \\ 0 & 0 & 0 & 0 & 1 & -1 \end{bmatrix}$$

giving $$F = A_2 \log (A_1 P).$$

In general F can be any sequence of such functions, including the exponential operator.

A consistent estimator of the covariance matrix of F is the $(u \times u)$ matrix

$$V_F = H[V(P)]H^{\mathrm{T}}$$

where $H = [\mathrm{d}F(x)/\mathrm{d}x|x = P]$. Thus for the logit model, for example, since $\mathrm{d} \log (A_1 P)/\mathrm{d}P = \mathrm{D}_G^{-1} A_1$ where D_G is the diagonal matrix with the elements of G on the diagonal, we have

$$V_F = A_2 D_G^{-1} A_1 \; V(P) A_1^{\mathrm{T}} D_G^{-1} A_2^{\mathrm{T}}.$$

The structural model can be written in the usual notation as

$$F(\pi) = X\beta$$

where X is the $(u \times t)$ design matrix and β the $(t \times 1)$ vector of unknown coefficients. Then the generalized least squares estimator of β which is BAN is given by

$$\hat{\beta} = (X^{\mathrm{T}} V_F^{-1} X)^{-1} X^{\mathrm{T}} V_F^{-1} F$$

with $$V(\hat{\beta}) = (X^{\mathrm{T}} V_F^{-1} X)^{-1}.$$

The goodness of fit statistic for the model is given by

$$Q = (F - X\hat{\beta})^{\mathrm{T}} V_F^{-1}(F - X\hat{\beta})$$

which is asymptotically distributed on the null hypothesis as χ^2 with $(u - t)$ degrees of freedom.

We can further test hypotheses about the elements of β in the usual form

$$H_0 : C\beta = 0$$

where C is a $(c \times t)$ matrix; with the test statistic

$$Q_c = (C\hat{\beta})^{\mathrm{T}}[C(X^{\mathrm{T}}V_F^{-1}X)^{-1}C^{\mathrm{T}}]^{-1}\, C\hat{\beta}$$

which has an approximate χ^2 distribution with c degrees of freedom. The predicted values of $F(\pi)$ are given by

$$\hat{F} = X\hat{\beta}$$

with $$V(\hat{F}) = X(X^{\mathrm{T}}V_F^{-1}X)^{-1}X^{\mathrm{T}}.$$

For the logit model in our example, we have

$$\operatorname{logit} p_k = \beta_0 + \beta_1 x_k$$

where $k = 1, 2, 3$ and $x_k = 0, 7, 11$

so that $$X = \begin{bmatrix} 1 & 0 \\ 1 & 7 \\ 1 & 11 \end{bmatrix}$$

and we obtain

$$\hat{\beta}^{\mathrm{T}} = (-0{\cdot}946, 0{\cdot}031)$$

$$V(\hat{\beta}) = \begin{bmatrix} 4{\cdot}808 & -0{\cdot}188 \\ -0{\cdot}188 & 0{\cdot}030 \end{bmatrix} \times 10^{-4}$$

So that the elements of $\hat{\beta}$ are highly significantly different from zero. The results from the different models are summarized in Table 4.16.

Koch *et al.* (1977) also show how to apply these methods to large tables with many occasions and some possibly zero cell frequencies, and Standish *et al.* (1978) deal with the problem of missing data.

References

Armitage, P. (1955). Tests for linear trends in proportions and frequencies, *Biometrics*, **11**, 375–386.

Armitage, P. (1971). "Statistical Methods in Medical Research", Blackwell, Oxford.

Atwood, C. L. and Taube, A. (1976). Estimating mean time to reach a milestone using retrospective data, *Biometrics*, **32**, 159–172.

Baker, R. J. and Nelder, J. A. (1978). "Generalised Linear Interactive Modelling Manual", Release 3, Numerical Algorithms Group, Oxford, England.

Battese, G. E. and Fuller, W. A. (1973). An unbiased response model for analysis of categorical data, *Proceedings of the Social Statistics Section of the American Statistical Association*, 202–207.

Bishop, Y. M. M., Feinberg, S. E. and Holland, P. W. (1975). "Discrete Multivariate Analysis", MIT Press, Cambridge, Mass.

Blalock, H. M. (ed) (1971). "Causal Models in the Social Sciences", Aldine–Atherton, Chicago.

Bock, R. D. (1976a). "Multivariate Statistical Methods in Behavioural Research", McGraw-Hill, New York.

Bock, R. D. (1976b). Basic issues in the measurement of change, *In* "Advances in Psychological and Educational Measurement", (Gruijter, D. N. M. and Van der Kamp, L. T. J. eds), Wiley, New York.

Campbell, D. T. (1963). From description to experimentation: interpreting trends as quasi-experiments, *In* "Problems in Measuring Change" (Harris, C. W., ed), University of Wisconsin Press, Madison.

Clayton, D.G. (1978). A model for association in bivariate life tables and its application in studies of familiar tendency in chronic disease incidence, *Biometrika*, **65**, 141–151.

Clayton, D. G. (1979). Retrospective studies of the aetiology of chronic diseases, *In* "The Recall Method in Social Surveys", (Moss, L. and Goldstein, H., eds) University of London Institute of Education, London.

Cochran, W. G. and Rubin, D. R. (1973). Controlling bias in observational studies: a review, *Sankhya*, **35**, 417–446.

Cornfield, J. and Haenszel, W. (1960). Some aspects of retrospective studies, *Journal of Chronic Disease*, **11**, 523–533.

Cox, D. R. (1972). Regression models and life tables, *Journal of the Royal Statistical Society*, B **34**, 187–220.

Fletcher, C., Peto, R., Tinker, C. and Speizer, F. E. (1976). "The Natural History of Chronic Bronchitis and Emphysema". Oxford University Press, London.

Fogelman, K. R. and Goldstein, H. (1976). Social factors associated with changes in educational attainment between 7 and 11 years of age, *Educational Studies*, **2**, 95–105.

Fuller, W. A. and Hidiroglou, M. A. (1978). Regression estimation after correcting for attenuation, *Journal of the American Statistical Association*, **73**, 99–104.

Galton, F. (1886). Regression towards mediocrity in heredity stature, *Journal of the Anthropological Institute*, **15**, 246–263.

Gleser, G. C., Cronbach, L. J. and Rajaratnam, N. (1965). Generalizability of scores influenced by multiple sources of variance, *Psychometrika*, **30**, 395–418.

Goodman, L. A. (1973). The analysis of multidimensional contingency tables when some variables are posterior to others: a modified path analysis approach, *Biometrika*, **60**, 179–192.

Goodman, L. A. (1974). The analysis of systems of qualitative variables when some of the variables are unobservable. Part I–A, Modified latent structure approach, *American Journal of Sociology*, **79**, 1179–1259.

Hannan, M. T. and Young, A. A. (1977). Estimation in panel models: results on pooling cross-sections and time series, *In* "Sociological Methodology", (Heise, D. R., ed) Jossey-Bass, San Francisco.

Healy, M. J. R. and Goldstein, H. (1978). Regression to the mean, *Annals of Human Biology*, **5**, 277–280.

Hills, M. (1974). "Statistics for Comparative Studies", Chapman and Hall, London.

James, K. E. (1973). Regression towards the mean in uncontrolled clinical studies, *Biometrics*, **29**, 121–130.

Johnston, J. (1972). "Econometric Methods", 2nd Edition, McGraw-Hill, New York.

Joreskog, K. G. (1970). A general method for analysis of covariance structures, *Biometrika*, **57**, 239–251.

Joreskog, K. G. and Sorbom, D. (1976). Statistical models and methods for analysis of longitudinal data, *In* "Latent Variables in Socio-economic Models", (Aigner, D. J. and Goldberger, A. S., ed) North Holland, Amsterdam.

Kadane, J. B., McGuire, T. W., Sanday, P. R. and Staelin, R. (1976). Models of environmental effects on the development of IQ, *Journal of Educational Statistics*, **1**, 181–231.

Kendall, M. G. and Stuart, A. (1967). "The Advanced Theory of Statistics", Vol. II, Griffin, London.

Kenny, D. A. (1973). Cross-lagged and synchronous common factors in panel data, *In*, "Structural Equation Models in the Social Sciences", (Goldberger, H. S. and Duncan, O. D., eds) Seminar Press, New York.

Koch, G. G., Imrey, P. B. and Reinfurt, D. W. (1972). Linear model analysis of categorical data with incomplete response vectors, *Biometrics*, **28**, 663–692.

Koch, G. G., Landis, J. R., Freeman, J. L., Freeman, D. H. and Lehnen, R. G. (1977). A general methodology for the analysis of experiments with repeated measurement of categorical data, *Biometrics*, **33**, 133–158.

Liddel, F. D. K., McDonald, J. C. and Thomas, D. C. (1977). Methods of Cohort Analysis: appraisal by application to asbestos mining. *Journal of the Royal Statistical Society*, A **140**, 469–491.

Lord, F. M. (1967). A paradox in the interpretation of group comparisons, *Psychological Bulletin*, **68**, 304–305.

Lord, F. M. and Novick, M. R. (1968). "Statistical Theories of Mental Test Scores", Addison-Wesley, Reading, Massachusetts.

Mantel, N. (1973). Synthetic retrospective studies and related topics, *Biometrics*, **29**, 479–486.

Moss, L. and Goldstein, H. (1979). "The Recall Method in Social Surveys", University of London, Institute of Education, London.

Mote, V. L. and Anderson, R. L. (1965). An investigation of the effect of misclassification on the properties of χ^2 tests in the analysis of categorical data, *Biometrika*, **52**, 95–109.

Nelder, J. A. (1974). Log-linear models for contingency tables: a generalization of classical least squares, *Applied Statistics*, **23**, 323–329.

Peto, R. (1973). Experimental survival curves for interval censored data, *Applied Statistics*, **22**, 86–91.

Prentice, R. L. (1976). Use of the logistic model in retrospective studies. *Biometrics*, **32**, 599–606.

Prentice, R. L. and Breslow, N. E. (1978). Retrospective studies and failure time models, *Biometrika*, **65**, 153–158.

Richardson, D. M. and Wu, D. (1970). Least squares and grouping method estimates in the errors-in-variables model, *Journal of the American Statistical Association*, **65**, 724–748.

Scheffé, H. (1959). "The Analysis of Variance", Wiley, New York.

Sorbom, D. (1976). A statistical model for the measurement of change in true scores, *In* "Advances in Psychological and Educational Measurement", (Gruijter, D. N. M. and van der Kamp, L. T. J., eds), Wiley, London.

Standish, W. M., Gillings, D. B. and Koch, G. G. (1978). An application of Multivariate ratio methods for the analysis of a longitudinal clinical trial with missing data, *Biometrics*, **34**, 305–317.

Swan, A. (1969). Computing maximum likelihood estimates for parameters of the normal distribution from grouped and censored data, *Applied Statistics*, **22**, 86–91.

Tanner, J. M., Whitehouse, R. H. and Takaishi, M. (1966). Standards from birth to maturity for height, weight, height velocity and weight velocity: British children, 1965, II. *Archives of Disease in Childhood*, **41**, 613–635.

Tukey, J. W. (1976). "Exploratory Data Analysis", Wiley, New York.

Warren, R. D., White, J. K. and Fuller, W. A. (1974). An errors-in-variables analysis of managerial role performance, *Journal of the American Statistical Association*, **69**, 886–893.

Werts, C. E. and Linn, R. L. (1970). A general linear model for studying growth, *Psychological Bulletin*, **73**, 17–22.

Zellen, A. (1962). An efficient method of estimating seemingly unrelated regressions and tests for aggregation bias, *Journal of the American Statistical Association*, **57**, 348–368.

6

Population Standards or Norms

The simplest population standards or norms describe how a measurement is distributed among the individual members of the population. They tell us what proportion of individual members in a well defined population have measurements below or above any chosen value. This then allows us to make judgements about whether any given individual is "atypical", that is whether the individual's measurement lies outside the range in which most measurements fall. Standards for combinations of measurements perform a similar function, and we shall discuss the special cases of velocity, or rate of change, standards and bivariate standards in a later section. To illustrate the basic principles, however, the initial discussion deals with single or univariate measurements only.

As an illustration Fig. 6.1 shows cross-sectional standards for height attained by British boys from birth to 19 years, with percentile curves locating the heights below which selected proportions of the population fall. (It also shows a certain type of longitudinal standard, which we shall discuss later.) The third and 97th percentiles are conventionally taken as the extreme positions for judging the typicality of a child but as will be seen later, there may be more objective criteria for selecting cut-off points if, indeed, a yes or no, typical or untypical, answer is the one required.

Six distinct issues have been touched on so far. These are: the definition and use of the word "population"; how to use the data collected to estimate the standards; how to present the standards so that they are readily usable; how they are to be used in making judgements; how to decide how many and which measurements should be used; and how to sample the population. This chapter will deal mainly with continuous measurements since the practical use of discrete measurements for standards is less widespread. We do, however, present a short discussion of standards for discrete measurements in Section 6.6.

The first four issues above are discussed in turn, the last two having been dealt with in Chapter 2. Firstly, we need to say why, in a book on longitudinal studies and the measurement of change, much of the space in this chapter is devoted to purely cross-sectional standards. To begin with these are by far the most common standards in use, as opposed to longitudinal standards which are based on measurements made on two or more separate occasions on each individual. Also, a large class of cross-sectional standards are concerned with development and are thus intrinsically connected to an age or time scale, so that they can be seen in the context of change.

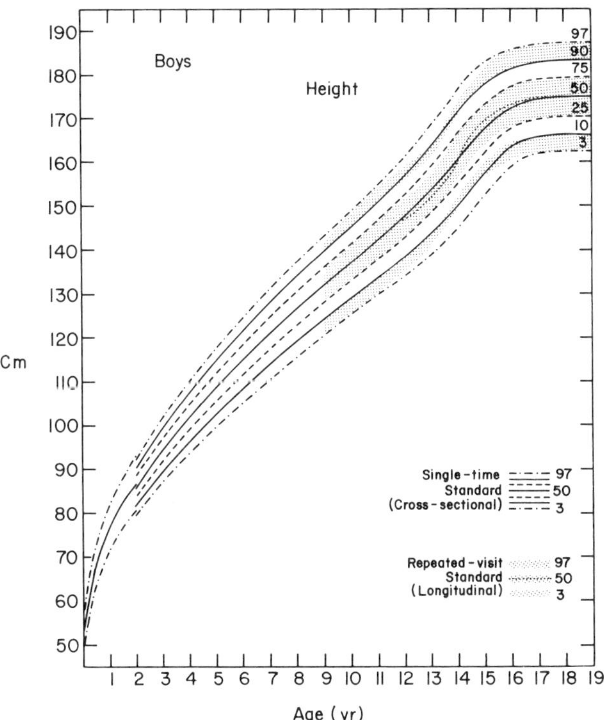

Fig. 6.1 British cross-sectional and adolescent individual-time-scale longitudinal standards for height of boys. (Tanner *et al.*, 1966)

6.1 Defining the Population

A population may be as large or as small as we wish. We could include every single human being or just those human beings born in one street. Most populations fall between these two in size, but geographical location is only one characteristic of a population. The time period during which the population exists must also be defined, and this raises the problem which was discussed at the beginning of Chapter 2. Time does not stand still and the characteristics of any population can only be described for a time which has passed; extrapolation to the present or the future always involves an element of judgement. For cross-sectional or short-term longitudinal standards this problem is less serious since we can plan to update the standards at intervals. Nevertheless, these intervals may be relatively long and updating may be relatively costly, so that we may have to rely on extrapolations from past standards. This will be particularly unsatisfactory where population characteristics are changing rapidly, such as is sometimes the case with physical growth standards in developing countries. Unfortunately it is also often the case that in such countries the available resources for carrying out standardizations are the most limited.

Any individual is simultaneously a member of many different populations. He may belong to the population of five-year-olds living in France, or to the population of five-year-old boys in Europe, etc. Likewise, when we consider longitudinal standards, the problem of the definition of a population over time, as discussed in Chapter 2, also arises. This raises two related problems. Firstly, we have to decide which populations should be considered when constructing standards. Secondly, we have to decide which population standards should be used when an individual belongs to more than one population for which standards are available. In general there are no clear answers to these questions since any answer will depend ultimately on the reasons for constructing and using the standards. The following discussion nevertheless outlines some of the issues involved.

We begin by describing one real population to illustrate the problems. This population consists of the children aged 7·0 years in Britain in the mid-1960s and we shall consider the estimation of cross-sectional height standards. Ignoring the slight vagueness in the definition of the time period, the simplest standards will be those based on the whole population. Alternatively we can divide this population into several sub-populations, defined, for example, by the parity or birth order of each child and quote separate standards for each.

Figure 6.2 shows three separate percentile standards for parity using data from the National Child Development Study (Goldstein, 1974). Also shown are standards for children from extreme social class groupings, Professional and Managerial Workers (I & II) and Unskilled Workers (V) and for two combinations of social class by parity groupings.

It is evident that, depending on which sub-population is used, the percentile position of a given child changes. Thus, using the total population, a sixth percentile child with a height of 110·0 cm becomes a third percentile child if he comes from Social Class I or II and is referred to the standards pertaining to that sub-population. It is, of course, quite reasonable to expect that the extra information obtained about the classification of an individual into a sub-population will cause us to modify our classification, and also any action we take as a result of that classification. If we wish to use the separate birth order standards we need to take into account their practical convenience and to decide whether the extra cost in compiling such standards and the extra information and care needed in using them is offset by the gain in discrimination. With some factors there is so little difference between separate sub-population standards that little is gained by presenting separate standards. This would be the case for sex at age seven years, in fact the main reason for presenting separate standards for boys and girls is because of the large differences which appear at puberty. Also, of course, a factor such as sex is readily determined whereas variables such as social class are more difficult to measure accurately. Above all, however, we need to consider the purposes for which the standards have been compiled and how these purposes are fulfilled by different sub-population classifications. Because this issue has arisen most acutely for growth standards in developing countries, we shall use these as an illustration.

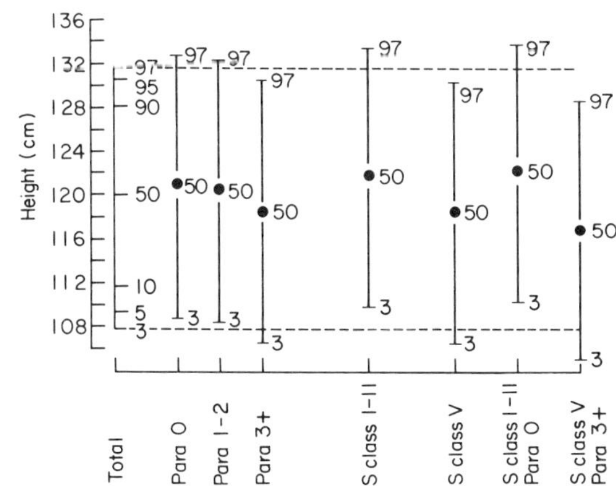

Fig. 6.2 Standards for height of British seven-year-olds. (National Child Development Study)

First of all we discuss the purely descriptive function of separate sub-population standards. Differences between sub-populations are usually described in terms of average values and standard deviations, or alternatively by specifying the complete percentile distributions. Such descriptions are not only of scientific interest, but also often of considerable practical importance since they may indicate those groups to whom extra social, medical or educational resources should be given. When we turn to the normative use of sub-population standards for making individual clinical judgements, then controversial issues arise. Should, for example, a member of an under-privileged social group be judged by standards pertaining to that group, or should he or she be judged by the standards of a privileged group? A child at the 20th percentile of height in the former group might only be at the third percentile in the latter, and it is not immediately obvious which judgement, if either, is appropriate.

One line of argument justifies the use of the developmental standards of the most privileged or élite group for all children in a society on the following grounds. Given genetic similarities between all the groups in a population it may be assumed that the less privileged individuals are not achieving their potential and it is this potential which the privileged standards represent. A slight variant of the argument recognizes that so-called élite groups may have passed the stage of optimum development, as, for example, when overnutrition leads to obesity. In this case an alternative optimum group is sought. The argument concludes that it is reasonable therefore to judge less privileged individuals against their assumed potential (International Union of Nutritional Sciences, 1972). Apart from the practical difficulty of ensuring genetic homogeneity, there is a considerable problem in deciding what the word "optimum" really means. What is optimal for one group in one environment may not be optimal for another group in another environment. A child in a developing country in a very poor environment, for example, receiving small amounts of energy from food, may have adapted his life style and his metabolism quite well to his environment, even though he is very small when judged by élite standards. This is, of course, no argument for neglecting the poverty of that child's environment. It does imply, however, that to treat the child in the same way as a very small child from an élite group, for example by special food supplementations, may not be in that child's best interests, nor make the best use of resources. It could spoil his adaptation to an otherwise unchanging environment, and divert whatever resources might be available away from an attempt to improve that environment. We shall not pursue this issue further, except to note that the arguments surrounding it are subtle ones and involve elements of both science and ideology.

6.2 Types of Measurement

Broadly speaking, there are two ways in which measurements can be used for the preparation of standards. In one, a measurement acts as a proxy for one or more other measurements and in the other the measurement itself is of direct interest. The former use is the most common, and an example is the use of height as an indicator of physical development. (Height, of course, may also be of interest in itself, for example to a manufacturer of children's clothes.) Since it is known that height is influenced by nutrition and disease, an atypical height can be used to justify a more detailed investigation of a child to detect abnormal nutrition or disease. If it were possible, however, to measure the nutritional state directly, then this would be an example of the latter use. A decision about what to measure will be influenced by the precise questions to be answered and the relative costs of different procedures. The use of proxy measurements as a screening device can be viewed as part of a two-stage procedure, whereby atypical individuals selected by the proxy are then assessed using a more direct measure. Further stages may be added and a number of different direct or proxy measures applied at different stages. The costs of the different procedures can be related both to the proportion of individuals correctly diagnosed, and to the proportion of individuals who have the characteristic looked for, but who have not been identified by the proxy measurement. These proportions can be altered by changing the value of the proxy variable beyond which individuals are selected for further study. This will then change the number of individuals selected and hence the cost of using the direct measure. Thus, the number of individuals finally identified can be related to total cost, or cost per individual, for different selection procedures. In order to make such estimates, we need to know the probability of having the characteristic for given values of the proxy variable. Unfortunately it will often be this information which is lacking.

Where the total resources for making measurements are fixed, so that only a certain number of individuals can be measured, an efficient procedure will consist of selecting those individuals with the most extreme values on the proxy measurement (see Goldstein (1972a) for a further discussion of screening procedures).

6.3 Age Changes and Time Trends

The need to understand secular, or time trends, including seasonal effects, was dealt with in Chapter 2. We now discuss a particular effect which can arise when constructing standards based on age. The effect of a time trend, if it is large, is illustrated in Fig. 6.3.

Two cohorts, one year apart in age, are measured at year 1 and year 2. The eight-year-olds of cohort 1 at year 2 are 1 unit larger than the eight-year-olds of cohort 2 at year 1, so that a time trend of 1 unit (for eight-year-olds) has taken place during one year. Also, the difference between the cohorts one year apart is 1 unit. The change for each cohort over one year of age is 2 units, this being the sum of the time trend and the cross-sectional age difference. Thus the cross-sectional age difference at a point in time is not equal to the longitudinal change over an equivalent number of years. In many populations secular trends are relatively small. For example, in Britain after the second world war, the secular trend in height amounted to not more than 2% of any cross-sectional age difference in pre-adolescent children. However, in rapidly developing societies, secular trends may become relatively large and care will then be necessary to allow for their effects and to design samples to measure their magnitude. Where the average age change, judged by longitudinal standards, is not the same as that given by cross-sectional standards, it may be necessary to plan frequent updatings of the standards.

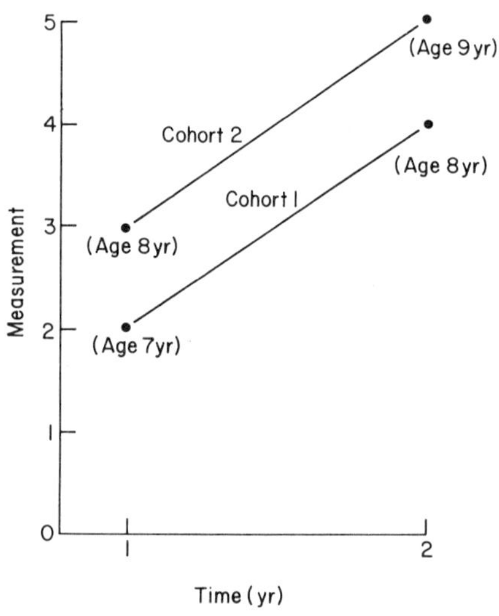

Fig. 6.3 Hypothetical illustration of age and time trend effects.

For some measurements, a special kind of time-trend effect is created by the nature of the circumstances in which the measurement is made. Such a situation may occur, for example, in an educational system, where children of different ages spanning, say, a single year, move together through the system. At the beginning of the year the oldest child has the same age which the youngest child will have one year later, but, on average, the two will not necessarily have the same measured attainments at this age. It has been found that, at some ages, the average longitudinal change over a year is greater than the average difference between children one year apart in age at a given point in time. The importance of this effect in mental testing is discussed in detail by Goldstein and Fogelman (1974).

6.4 Longitudinal Standards for Two Occasions

The simplest type of longitudinal standard is based on a measurement made, on two separate occasions, on the same individual. Many problems can usefully be discussed for this case before treating the general case of several occasions, and, in fact, almost the only longitudinal standards in common use are the so-called "velocity" standards, which deal with the average rate of change in a measurement between two occasions. The occasions are usually chosen one year apart to avoid seasonal effects and the velocity (or rate of change) is plotted at the age midway between the two ages on which the calculation is based. However, Marshall (1971) does give height velocity standards for periods of less than one year, which take into account seasonal variation.

Velocity standards, constructed on the basis of a fixed time interval between occasions, have the drawback that they can be used only where measurements are also made with the same time interval between them. A velocity measured over a shorter or longer period will not only encounter the problem of seasonal change, but even where the average velocity agrees with that of fixed-time-interval standards, the variability, and therefore the percentile estimates, will not agree. For example, the standard deviation of height velocity over a two-year interval for height between five and ten years is only about 60% of that for a one-year interval. If deviations from the predetermined interval are not too large, however, then errors will be small. There appear to be no published data which indicate how large a deviation might be acceptable in practice.

Problems of sampling and estimation, which arise in the construction of velocity standards, are similar to those for distance standards. By regarding each velocity measurement as a single measurement taken at the mid point of the age interval, we may apply the techniques discussed in Chapter 2. For example, the rate of change of height velocity with age (height acceleration) is greatest for boys for the age ranges 0–2 and 13–17 so that the sampling will be concentrated in these ranges (Tanner *et al.*, 1966). In constructing velocity standards, as pointed out above, we need to ensure that all pairs of measurements cover the same age interval, apart from minor fluctuations. Where individuals are actually measured on more than two occasions it may be possible to adjust the measurements to enclose the required age interval, and a technique for doing this is described in Chapter 7.

Although the difference, or rate of change, in a measurement on two occasions is a simple and convenient summary of the two measurements, it is not necessarily the most appropriate, nor does it necessarily make the best use of the information available. It is a "symmetrical" function of the two measurements which gives the same status to each measurement, whereas, as we saw in Chapter 5, a conditional approach which takes account of the direction of time may be more useful. We shall consider conditional standards in a later section.

6.5 Longitudinal Standards for Several Occasions

Once we move beyond consideration of two occasions, the problem of the appropriate function of measurements to use for longitudinal standards becomes much more complicated. We may extend, for example, the idea of velocity measurements, to the calculation of differences between successive velocities, so as to give acceleration measurements. More generally, however, we will be concerned with the pattern of growth, as described by the fitting of growth curves, and the use of the estimated parameters of these curves to characterise individuals. For short time intervals we might attempt to fit polynomial curves, or alternatively we might study the shape of the observed growth curves to see if useful regularities can be found. The variation in particular patterns may then be used as the basis for growth standards. One such standard has been devised by Tanner *et al.* (1966) for growth in height and weight during adolescence, the characteristic shape of adolescent growth being determined from longitudinal data over a wide age range (see Fig. 6.1). Terada and Hoshi (1965) describe a preliminary exploration of physical growth patterns in the first three years of life.

If polynomial curves are fitted to longitudinal data, we can use the method proposed by Bowden and Steinhart (1973) to construct percentile bands which include a given percentage of such growth curves. For those stages of growth where a low order polynomial is a good approximation, and, for measurements with approximately normal distributions, this approach should provide a good estimate of the range of variation in growth patterns.

It is sometimes suggested that longitudinal data should be plotted on cross-sectional standards, but, in the absence of information about expected variability in longitudinal patterns, it is difficult to see how any observed patterns can be interpreted reliably. For example, in a long sequence of measurements, the fact that a few lie beyond some of the more extreme cross-sectional percentiles may not imply an abnormal developmental pattern. It is only by constructing true longitudinal standards, such as described above, that we can reliably detect abnormal patterns.

6.6 Standards for Discrete Measurements

For many aspects of development it is not possible to make measurements on a continuous scale, either because we do not have fine enough measuring instruments, for example to measure the continuous process of dental calcification (see Chapter 3), or because what occurs is itself a discrete event, such as menarche. In the former case we may be able to obtain a continuous scale by assigning scores to stages, as described in Chapter 3. In the following discussions we shall use ratings of pubertal development as one example of a discrete measurement.

6.6.1 Cross-sectional Standards

According to Tanner (1962) during puberty children go through five recognizable stages of pubic hair, genitalia and, for girls, breast development. Figure 6.4, from Van Wieringen *et al.* (1971), shows the cumulative proportion of girls who are at or beyond each stage of breast development at different ages. Thus, at the age of 16·0 years about 95% have reached stage 4 so that a girl of this age who has not yet reached stage 4 might be regarded as atypical. We note that it is no longer possible to present estimates for all possible percentiles at any given age, since the discrete nature of the measurement determines the percentiles which can actually be observed.

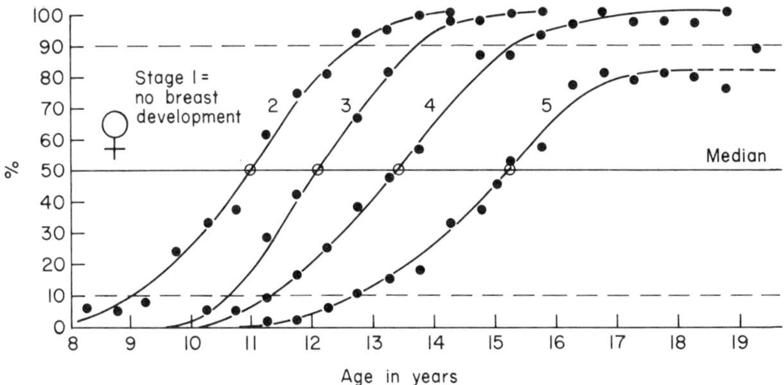

Fig. 6.4 Cumulative frequency distribution of breast development stages. (Van Wieringen *et al.*, 1971)

We can use two somewhat different approaches in order to estimate the cumulative percentages or percentiles. If we consider each adjacent pair of stages separately, then at each age we can estimate the percentage who have passed the transition between stages, in the same way as for an event such as menarche. Methods for doing this, based on a logit transformation, are described in Chapter 5. Alternatively, we can regard the set of percentages in each of the 5 stages as a (multivariate) dependent variable, which is to be related to age, giving us estimated percentages in each stage for each age. Thus, for example, if at a given age only 3% of boys were in the first stage of pubic hair, then any boy of this age observed to be in this stage would be judged to be at or below the third percentile of pubic hair development. This latter approach is not only the direct analogue for discrete data of the method used for continuous data, it also uses the available information more efficiently than the former method. The technique for estimating the percentages is described also in Chapter 5. Sampling strategies for the efficient estimation of percentiles are discussed in the section on developmental standards in Chapter 2.

6.6.2 Longitudinal Standards

Since longitudinal standards for discrete measurements do not seem to exist yet, the following discussion will use a hypothetical example of a 3-stage rating. By analogy with velocity standards, the simplest 2-occasion standards will give the percentages of individuals in each combination of stages at the two occasions. For a 3-stage rating there will be six such combinations, assuming that transitions from higher to lower stages do not occur. As with continuous measurements, we restrict the difference between the two occasions to be a fixed interval, say one year, and plot the percentages in each stage combination at the mid point of the interval. Because transitions between stages are from lower to higher stages, however, it seems more natural to adopt conditional standards where we study the percentages of individuals in stages at occasion 2 for given stages at occasion 1, this being the discrete measurement analogy of conditional standards for continuous measurements, discussed below.

When we move to three or more occasions, the situation becomes more complicated and we might consider, say for three occasions 1 year apart, the probability of being in each stage at occasion 3, for given combinations of stages at occasions 1 and 2.

6.7 Conditional Standards

We consider now the construction and use of conditional standards for continuous measurements.

In a general sense, conditional standards consist of the percentile distribution of a measurement for a specified set of values of one or more other measurements. The commonly used standards of birthweight for length of gestation (Tanner and Thompson, 1970) are conditional standards in this sense, as are those for children's height, given the heights of their parents (Tanner *et al.*, 1970).

The term "conditional standards" is also used in a somewhat narrower sense to refer to a method of constructing longitudinal standards (Cameron, 1978). In the two-occasion case, instead of calculating the distribution of the velocity, the percentile distribution of the second occasion measurement is determined for every possible value of the first occasion measurement. In the simplest case, the mean of the second occasion measurement is linearly related to the first occasion measurement and the distribution about the mean is determined. Thus, where the distribution is normal, a simple linear regression of second occasion measurement on first occasion measurement will provide all the necessary information for constructing the percentiles. The 50th percentile is simply the value predicted by the regression equation and the standard deviation about this line is $\sigma_2(1 - \rho^2)^{\frac{1}{2}}$, where σ_2^2 is the second occasion measurement variance and ρ is the correlation between occasions. A detailed study of longitudinal conditional standards for height is given by Cameron (1978). Where the relationship is more complex, we may still estimate the percentile distribution of the second occasion measurement for given first occasion measurements, using a procedure similar to that described for constructing standards over a wide age range, with the first occasion measurement taking the place of age. (Goldstein, 1972b). The extension to three or more occasions involves, for combinations of measurement values at all previous occasions, the estimation of the percentile distribution at the final occasion.

The use of longitudinal conditional standards makes explicit use of the direction of time, and allows a more sensitive judgement than that offered by purely cross-sectional standards. Nevertheless, the measurement which is conditioned upon, that is the first-occasion measurement in the two-occasion case, may also provide useful information in its own right, and this should not be ignored. Like the velocity standards, longitudinal conditional standards provide a particular one-dimensional summary of a two (or more) dimensional set of measurements. This leads us to the final section of this chapter which deals with bivariate standards.

6.8 Bivariate Standards

Until now we have described only single-measurement or "univariate" standards. In the case of velocity and longitudinal conditional standards, we have made implicit assumptions about which (single valued) function of two (or more) measurements at different occasions should be used. If such prior assumptions are not made, then we must retain the measurements at each occasion. Thus for the two-occasion case we are led to a study of the "bivariate" or joint distribution of the two measurements. Figure 6.5 represents a bivariate normal distribution of two measurements. As in the case of univariate standards, we wish to select a region of extreme values within which we may refer to individuals as atypical.

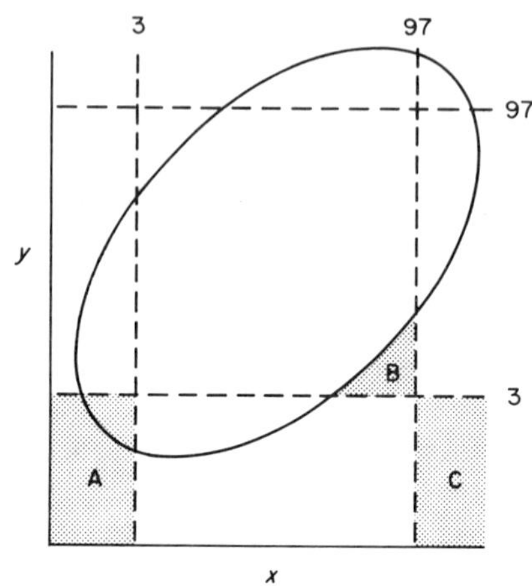

Fig. 6.5 A bivariate normal probability distribution showing equal probability 97th percentile ellipse and univariate third and 97th percentiles

$(\sigma_X = \sigma_Y, \rho = 0{\cdot}5)$

In Fig. 6.5 the dotted lines represent the separate univariate third and 97th percentiles for the two measurements X and Y, and the ellipse represents the "equal probability" contour enclosing 97% of the population. Such a contour is defined so that the probability of a randomly chosen individual occurring anywhere inside the ellipse is greater than the probability of such an individual occurring at any point outside the ellipse. In the absence of other considerations, this definition of typical and atypical regions may be a reasonable one to adopt. In defining standards, however, we are not solely concerned with *any* atypical measurement, but with one which is atypical in a relevant manner. For example, when using univariate height standards we may be interested only in individuals who are too small, and the same consideration applies to bivariate standards. Suppose we are interested in individuals who are small on both occasions; that is, who fall into an extreme part of the distribution containing, say, 3% of individuals. There is no natural way of modifying the region defined by the ellipse in order to achieve this. The region in shaded area A clearly does contain such individuals, but it is defined by the univariate rather than the bivariate standards. Furthermore, the bivariate ellipse classifies some individuals as atypical, who are classified by both univariate standards (shaded area B) as typical, and classifies some individuals as typical, who are classified by both univariate standards as atypical. Furthermore, the atypical region defined by the ellipse pays no attention to the direction of atypicality. Thus we see that the simple device of enclosing a region with an equal probability contour is inadequate, and we are left with the problem of choosing an atypical region in accordance with the purpose for which the standards are to be used. In some applications it might be possible to specify such a region *a priori*. Region A, for example, only classifies as atypical those who are separately classified as such by each univariate standard. (This will include less than 3% of individuals, and we would need to alter the univariate percentiles to produce 3% of individuals in this region.) If we were interested in individuals with a large X measurement and a small Y measurement, we might choose a region similar to region C in Fig. 6.5, and so on. The increased choice of region means that bivariate standards may be used more sensitively to detect specific conditions associated with the measurements, than either of the corresponding univariate standards alone.

With univariate standards, the choice of an atypical region is usually straightforward in terms of extreme values, the problem consisting of which percentiles to choose as cut-off points. Where such standards are used in a clinical way, in order to screen out individuals, it will often be the case that more than one underlying abnormality can give rise to an observed extreme value. For example, both malnutrition and hormonal deficiencies can result in abnormally short children. When we come to bivariate standards, however, it will generally be the case that the separate measurements are differentially affected by different underlying abnormalities, so that different atypical regions will be optimum for such different abnormalities. We see, therefore, that the construction of satisfactory bivariate standards requires more empirical knowledge of the relationships between the measurements used and underlying abnormalities. In practice, also, we may be able to use one set of standards with a single atypical region to screen for several abnormalities if we find that the relationships are similar enough to be pooled. The following example shows how this approach to the construction of bivariate standards results in a considerable improvement over the use of a conditional standard. Although the measurements are not longitudinal, the same general principles apply in that case.

In Fig. 6.6, using data from the 1958 Perinatal Mortality Survey (Goldstein and Peckham, 1976) the two measurements used to construct bivariate standards are birthweight and length of gestation. The conditional percentiles of birthweight for given length of gestation are shown by the broken line, the fifth, 50th and 95th percentiles being illustrated. If neonatal mortality is used as the variable which the standards are designed to detect, then we need to estimate a neonatal mortality rate for each combination of birthweight and gestation. The continuous lines in the figure are formed by joining points with equal mortality rates, and are thus contours of constant mortality. So, we can define an atypical region as, say, the region below the 200 (twice average) contour, and this contains about 6% of all births. We see that such a classification selects a quite different set of births from that defined by the sixth percentile of the conditional, birthweight for length of gestation, standards.

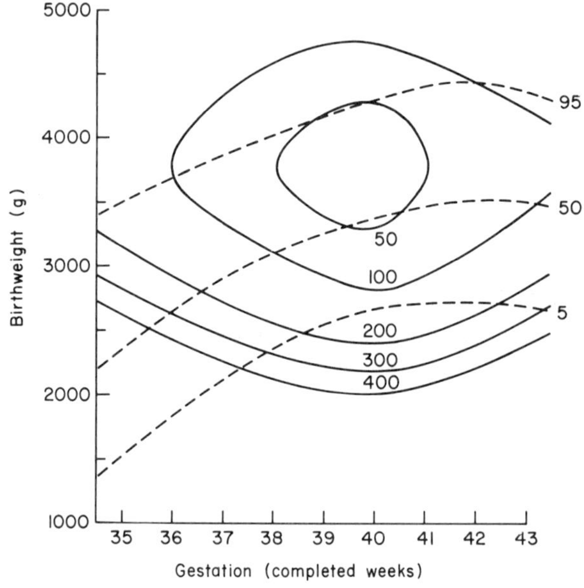

Fig. 6.6 Neonatal mortality rates by birthweight and gestation (1958 Perinatal Mortality Survey) mean rate = 100.

In fact, standards based on birthweight alone perform almost as well as the bivariate standards (for this range of gestation lengths). Moreover, Goldstein and Peckham show that birthweight alone is almost optimal when birthweight and gestation are used to predict subsequent educational and physical development. Thus if we wished to have a set of standards with a single atypical region to reflect abnormalities in neonatal mortality as well as in subsequent development, a reasonable compromise would be to use birthweight alone.

References

Bowden, D. C. and Steinhart, R. K. (1973). Tolerance bands for growth curves, *Biometrics*, **29**, 361–371.

Cameron, N. (1978). "An Analysis of the growth of London Schoolchildren". PhD thesis, University of London.

Demirjian, A. and Goldstein, H. (1976). New systems for dental maturity based on seven and four teeth, *Annals of Human Biology*, **3**, 411–421.

Goldstein, H. (1972a). The allocation of resources in population screening: a decision theory model. *Biometrics*, **28**, 499–518.

Goldstein, H. (1972b). The construction of standards for measurements subject to growth, *Human Biology*, **44**, 255–261.

Goldstein, H. (1974). Some statistical considerations on the use of anthropometry to assess nutritional status, *In* "Nutrition and Malnutrition". (Roche, A. F. and Falkner, F., eds) Plenum Press, New York.

Goldstein, H. and Fogelman, K. R. (1976). Age standardisation and seasonal effects in mental testing, *British Journal of Educational Psychology*, **44**, 109–115.

Goldstein, H. and Peckham, C. (1976). Birthweight, gestation, neonatal mortality and child development, *In* "The Biology of Human Fetal Growth". (Roberts, D.F. and Thomson, A. M. eds) Taylor and Francis, London.

International Union of Nutritional Sciences (1972). The creation of growth standards: a committee report of a meeting in Tunis, February 1971, *American Journal of Clinical Nutrition*, **25**, 218–220.

Marshall, W. A. M. (1971). Evaluation of growth rate in height over periods of less than one year, *Archives of Disease in Childhood*, **46**, 414–420.

Tanner, J. M. (1962). "Growth at Adolescence", Blackwell, Oxford.

Tanner, J. M., and Thompson, A. M. (1970). Standards for birthweight at gestation periods from 32 to 42 weeks allowing for maternal height and weight, *Archives of Disease in Childhood*, **54**, 566–569.

Tanner, J. M., Whitehouse, R. H. and Takaishi, M. (1966). Standards from birth to maturity for height, weight, height velocity and weight velocity: British children 1965, II, *Archives of Disease in Childhood*, **41**, 613–635.

Tanner, J. M., Goldstein, H. and Whitehouse, R. H. (1970). Standards for children's height at ages 2–9 allowing for height of parents, *Archives of Disease in Childhood*, **45**, 755–762.

Terada, H. and Hoshi, H. (1965). Longitudinal study of the physical growth in Japanese; (2) Growth in stature and bodyweight during the first three years of life, *Acta Anatomica Nipponica*, **40**, 116–123.

Van Wieringen, J. C., Wafelbakker, F., Verbrugge, M. P. and deHaas, J. H. (1971). "Growth Diagrams 1965 Netherlands", Wolters-Noordof Publishing, Groningen.

7
Data Processing

Since the early 1960s the availability of computers has encouraged the development of general purpose programs and systems for the processing of information collected by surveys. Such programs accept information obtained from the units or individuals in a survey and are designed to produce suitable summary tables, graphical displays and statistical analyses.

Despite a wide proliferation, very few survey programs can handle anything more complicated than simply structured cross-sectional data. They assume that each individual in the survey is associated with a single record containing a constant number of variables. Some programs will allow a differing number of variables for each record, but very few of these can cope adequately with longitudinal data, where each individual has a record for each measurement occasion, and where the number and type of measurements may change between occasions. For a longitudinal study, whether it is strictly a survey, or for example a designed experiment, we may readily adapt such a cross-sectional program so long as the number of occasions is very small. This is done by forming a master-record of which the separate occasion records are a part. Each variable in this new record is classified by both the measurement it represents and the occasion on which the measurement was made. Longitudinal relationships may then be specified explicitly by manipulating the variables labelled in this way. As the number of occasions increases, however, such a procedure becomes unwieldy since we will now need to carry out complex operations such as, for example, searching the data for particular longitudinal patterns. The master-record approach will not be discussed any further.

There is a similarity between longitudinal data and the "household" data structure. In the latter, a record may be obtained on each member of a household, with a variable number of such members, who will have some but not all information in common. Where the analysis of these records involves relationships among the members of the household, for example looking at differences in ages, we have similar problems to those described below. There are certain data base management systems which are designed to handle such hierarchically structured data. These programs recognize that an individual's record may contain a hierarchy of levels. With longitudinal data the first level will consist of those data common to all occasions, such as an individual's sex, date of birth, etc., and the second level will consist of the data for each occasion. Thus any such program which has good general facilities for manipulating data within and between levels should be capable of processing longitudinal data satisfactorily. For a description of one such program the reader is referred to Robinson *et al.* (1977).

In the following sections we describe in detail the special data processing requirements of longitudinal data, applicable both to general data base management systems and to special purpose longitudinal programs. First, we describe the ways in which any general purpose survey program receives, stores, manipulates and presents data.

7.1 General Purpose Survey Analysis Programs

A typical survey analysis system will accept data on individuals, check, transform and combine variables, store the edited data records on a magnetic tape (or disc) file and then carry out analyses and produce output.

The first operation is to read the input data, usually on punched cards. In some surveys the number of variables measured on each individual (or at each occasion for a longitudinal survey) is not fixed. Although the hierarchical data base management systems mentioned above will normally cater for this case, in order to keep the following discussion as simple as possible, we shall assume that this complication does not occur. If it does occur we can assume that it is dealt with satisfactorily by the system, since it does not raise any essentially new problems for longitudinal studies. This can be seen by considering one method of handling this case which consists of allowing for the maximum possible number of variables for each individual record and filling out the missing variables with special blank codes. The resulting records are then of fixed length and can be handled as straightforwardly as in the case of fixed length records where some variables for some occasions happen to be missing. The manner in which the program actually treats the data should not introduce special new problems.

Data are read into the central store of the computer, one individual's record at a time, and the various operations to check, transform and combine variables then take place. The program usually reads in a set of control cards in which are punched the appropriate instructions for carrying out these operations. Particular ways in which these operate will be outlined in a later section.

Possible errors in the data will be listed so that, after inspection, corrections can be made to the record file. Correcting the file is straightforward, but deciding which of the listed values are errors is not. A very large value, listed, say, by the checking program above the 99th percentile, might be either a mistake or truly an extreme value such as will occur on average in 1% of individuals. A decision as to whether it is incorrect needs to be justified on grounds other than its extreme value. In some cases the correct value can be obtained, for example where it is clear that a card has been mispunched. In other cases a measurement may be known to be incorrect but the correct value is unknown and in this case a code for "value unknown" will be inserted. There will still remain some values for which no clear decision can be reached and in these cases an element of judgement must be used. Removing too many extreme values artificially decreases the variability; not removing enough leaves the variability too large and may introduce serious bias.

A file-updating program or routine is required when further individuals enter a survey or, as in a longitudinal study, further data are collected on existing individuals, and when corrections are inserted.

Finally, there are programs or routines which read the edited data file, carry out analyses and produce output. Further manipulation of the data may be needed at this stage. Typically such programs calculate estimates of the parameters of continuous distributions (means, standard deviations, covariance matrices, etc.) and produce contingency tables.

We shall describe the basic operations as they apply to the data processing of longitudinal studies; we shall concentrate on those which have special application to such studies, and in particular those concerned with setting up, correcting and maintaining the record file. A more detailed description of the operation of survey analysis programs is given by Yates (1973).

7.2 Data Checking

Where a variable has a well-defined range (for example a score between 1 and 9), it is simple to check for illegal values. Similarly, where the value of one variable is logically dependent on the value of another, checks can readily be specified. If a girl, for example, is recorded as having passed menarche on a particular occasion, then she should also be recorded as having done so on subsequent occasions. In cross-sectional data we pay attention to extreme values, because those which are incorrect will tend to disturb our analyses more than non-extreme errors. In longitudinal data also we regard extreme values as suspicious but in addition we can judge the distribution of a variable in relation to its value or the value of another variable on the same individual on other occasions. In general this allows us to set much narrower limits and to obtain more sensitive procedures for defining extreme values.

For continuous variables we can use regression type equations relating measurements between an earlier and later occasion to judge the typicality of individuals on the later occasion. These will need to take account of possibly varying intervals between occasions, missed occasions, etc. The distribution of residuals about the regression line can be used to set approximate limits. Likewise we may use the association between discrete variables to set limits in that case, and we may extend the possibilities by considering longitudinal checks involving several variables at a time. Such checks require a knowledge of the relationships between measurements at different occasions and this will sometimes not be available prior to the study taking place. It may indeed be the purpose of the study to estimate such relationships, in which case we need to take care when editing data on this basis. Often, however, some information about relationships is available from other, perhaps pilot, studies. Where we do wish to determine the relationships using the data being collected by the study, editing could be postponed until data had been accumulated to determine the required relationships with sufficient accuracy. One such scheme might accumulate data for the first, say, 50 cases, estimate the required relationships, edit the first 50 cases, and then edit each new case as it arrived and at the same time update the estimated relationships with the new values. An alternative would be to wait until all the data had been collected before estimating the relationships. As well as checking currently collected data against previously collected data, we may also wish to recheck previous data against current data. Any corrections to previous data made as a result of this, if extensive, may suggest a reanalysis of any existing analysis based on the previous data.

We should always remember that the increased number of actual errors detected from having many checks has to be balanced against the labour of studying the increased number of suspicious values which are listed as possible errors. There will normally be an upper limit to the number of true errors detected as we increase the number of checks made. For any given set of data the optimum type and number of checking procedures will usually be found from experience. For physical growth data, for example, the relationships between pairs of occasions in the values of variables taken singly, together with pairs of variables compared cross-sectionally, should identify a sufficiently large number of errors (Goldstein, 1970). Naus *et al.* (1972) discuss the problem of checking from a general theoretical standpoint.

7.3 Transformations of Measurements

Above all else a longitudinal survey program must be able to manipulate variables across occasions easily, and to do this with simple instructions. New derived variables created in this way may be added to the basic record and stored along with other variables, or may be created for a particular analysis only. The following discussion will assume that the derived variables are to be preserved in the record file, but no new concepts are involved where the variables are to be used for a particular analysis only.

The derived variables could be stored in a separate file from the input data and linked by suitable identifiers. Generally, however, it will be preferable to store these variables along with the input data, each one being anchored to a particular occasion. For example, we might wish to define a height velocity variable which is the rate of change of height between consecutive pairs of occasions, and a natural choice for the anchoring occasion will be the second one of the pair. We might also wish to anchor a set of height measurements for surrounding occasions to a current occasion for subsequent manipulation, especially if a number of different calculations were to be performed on them. Thus at age t we might also have the measurements made at $t - 1$ and $t + 1$ (see next section).

The basic instruction which transfers variables, or longitudinal combinations of variables, onto a single chosen occasion, must be able to identify the occasions it will be using. It might do this explicitly, for example by referring to the first occasion. Generally, however, instructions must apply to all occasions satisfying specified general conditions. The main way of doing this is by referring to the difference in time, age or other scale between the chosen occasion and the other occasions which are to be related to it. For example, we might want to select all the occasions occurring between 0·9 and 1·1 years prior to the given occasion in order to calculate a velocity measurement over an approximate one-year period, and the program will then search an individual's record for such occasions. Such an instruction may often be needed for all the occasions in the record considered in turn. We might also use a variable other than age to define the occasion. For example, the previous occasion may be defined as that on which a particular test score differed from its current value by a given amount. We may also wish to satisfy more than one condition simultaneously, perhaps requiring that the change in a test score, as well as the difference in age, is within given limits. Finally, cross-sectional conditions may also be used, for example that the individual should be a female.

7.4 Variability of Occasion Times

We have seen, in Chapters 2 and 4, that longitudinal studies do not require every individual to be measured on every occasion, nor that individuals need to be measured at exactly predetermined times. Nevertheless, for some purposes, such as the construction of certain growth standards, it is convenient to have every individual measured at exactly the target time or within narrow limits of the target time. This may be a problem when individually based time scales, such as bone age, are used. Hence in this section we present one method of meeting this requirement for repeated measurements made on a continuous scale.

Figure 7.1 shows height measurements for a single child at three ages; 7·0, 7·8 and 9·0 years. The target occasion for the second measurement is 8·0 years and we require an estimate of the height at this age.

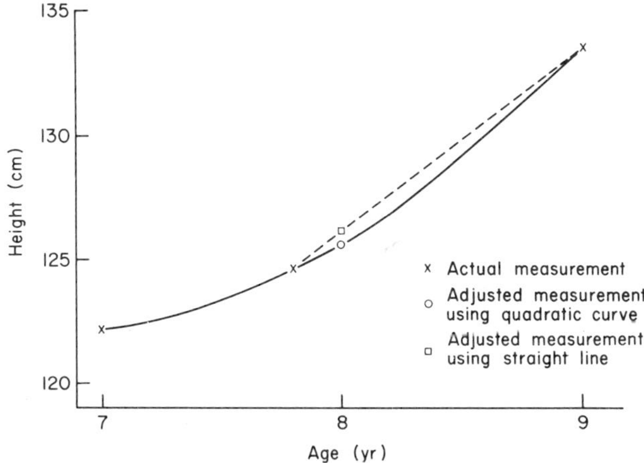

Fig. 7.1 Adjustment of height measurement made at 7·8 yr to 8·0 yr.

A quadratic curve has been fitted to pass exactly through the three measurement points, and the required adjusted value is the height read from this curve at age 8·0. Although we have chosen a polynomial curve, this is mainly for convenience and any three-parameter curve can be made to pass exactly through the three points. If a straight line joining the second pair of points were to be used, then we would obtain an alternative adjusted value. It is clear that the adjustment procedure depends on the form of the curve, which points are chosen, and how many points are used. Figure 7.1 shows that the straight line is not symmetric with respect to the occasion of interest and, in fact, measures only the average growth velocity after age 7·8. The quadratic curve is symmetric but it uses only three points and by including further points we might be able to obtain a better estimate. However, although too few points may give unreliable or biased estimates, if we use too many over a long time interval we may be imposing too much long-term regularity on the very changes we are trying to estimate. By restricting the time interval we make an assumption only of local regularity. In addition it is important that the time is small through which a measurement is adjusted and that the time interval spanned by the points used to fit the curve is not too large.

Experience with particular measurements will determine the best methods of making these adjustments. With physical growth data quadratic adjustment seems to be adequate, the adjustment being restricted to where the time difference between the measurement and the target occasion is less than about 20% of the target interval. Difficulties arise with the initial and final occasions since a quadratic curve which is symmetrical about the target occasion cannot be fitted. Strictly speaking, therefore, these points should be excluded although we could try using an asymmetric adjustment, possibly with a higher order polynomial. The program instructions must specify the order of polynomial to be fitted, the maximum time over which a measurement can be moved to its target time and a maximum time range for the set of points used to fit the curve. The adjusted measurement will then become a transformed variable.

To evaluate these procedures properly, studies are needed in which measurements are made both at the target times and at nearby times. Adjusted values can then be compared to the actual measured values. In the absence of such data these procedures should be used with caution, and the results examined carefully.

7.5 Creating and Updating the Data File

An individual's record consists of a set of measurement occasions each of which contains a set of variables (either raw measurements or transformed variables). Normally the occasions will be ordered along the time or age dimension and the records will be stored in this form by the computer on magnetic tape or disc, along with various items of bookkeeping information, such as the number of occasions for each individual, the number of measurements at each occasion, etc.

During the course of a longitudinal study, the file of completed occasions is corrected as errors are found and new occasions added. In some surveys extra data relevant to past occasions are collated with the stored data. None of these procedures raises new concepts and, since their functions are fairly obvious, they only justify a brief description of each.

After checking the data as described previously and deciding which corrections are necessary, we need programs or routines to alter values in individual records. This may involve changing one or more values on any occasion, and may also involve the values of transformed variables which have been derived from variables being corrected.

To update the data file with new occasions we need programs or routines which will add one or more occasions to an individual's record, and carry out transformations and edit checks as specified for previous data.

Further information on occasions arises in a number of ways, for example from merging information about physical and psychological development on the same individuals which was previously maintained on separate files. A program which collates such data should have facilities for combining information from the different files both cross-sectionally and longitudinally.

Figure 7.2 is a flow diagram which shows the procedures from the initial reading of records to final output. The role of the updating, correcting and collating programs or routines can be seen clearly from this.

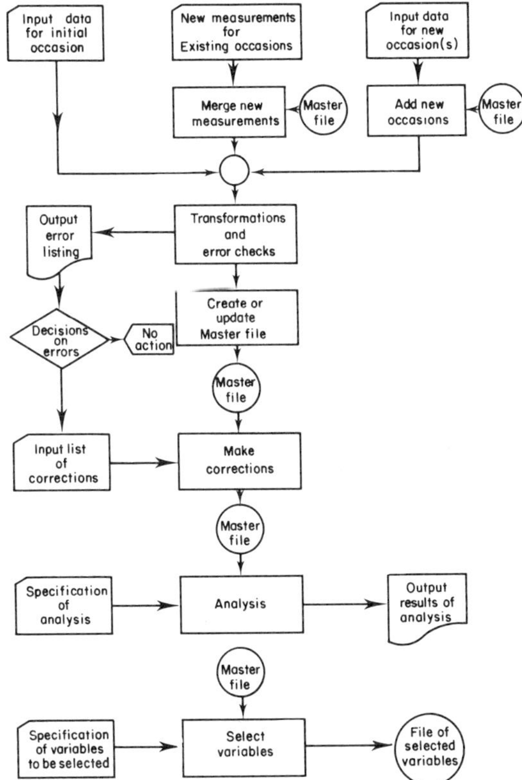

Fig. 7.2 Flow diagram for longitudinal survey system.

7.6 Output and Analysis

As far as this chapter is concerned, the end result of a survey analysis system is to produce data in a form to which the methods of analysis described in previous chapters can be applied. Output from the system may also consist of relatively simple summary statistics, and further analyses may use programs of a general nature not limited to longitudinal data.

The output program reads each individual record in turn from the file, carries out specified manipulations and produces selected data in the required form. The program may operate on raw or derived data. Useful summary statistics would include means, standard deviations, covariance and correlation matrices, contingency tables and estimates of percentiles. In addition, there will be facilities for creating a new file in a standard format containing selected variables for input to further analyses. For example, we may wish to study the interrelationships between the parameters of polynomial growth curves fitted to a given age range and this may be done conveniently by creating a file, each record of which contains the values of these parameters. Instructions for controlling these operations may be general or specific. They may refer to a particular occasion defined by the value of a variable, or to all the occasions which satisfy one or more conditions. An example of the latter would be an instruction to select, for each individual, those occasions where a transition had been made into a puberty stage from the previous stage, where the previous stage had been observed at an occasion which was one year earlier.

Finally, we should not forget the importance of clear and informative presentation of results. This is important enough in cross-sectional programs, and the increased complexity of typical output from longitudinal programs makes it even more desirable. Clear labelling and descriptions of how data are organised are essential. There should also be facilities for graphical output such as scatterplots, normal probability plots, plots of individual developmental patterns and so forth.

References

Goldstein, H. (1970). Data processing for longitudinal studies, *Applied Statistics*, **19**, 145–151.

Naus, J. I., Johnston, T. G. and Montalvo, R. (1972). A probabilistic model for identifying errors in data editing, *Journal of the American Statistical Association*, **67**, 943–950.

Robinson, B. N., Anderson, G. D., Cohen, E. and Gazdzik, W. F. (1977). Scientific Information Retrieval; Users' Manual S.I.R. Inc. Evanston, Illinois, U.S.A.

Yates, F. (1973). The analysis of surveys on computer–features of the Rothamsted survey program, *Applied Statistics*, **22**, 161–171.

Index